New Developments in Quantitative Trading and Investment

Series editors

Christian Dunis
Liverpool John Moores University
Liverpool, UK

Hans-Jörg von Mettenheim
Leibniz Universität Hannover
Hannover, Germany

Frank McGroarty
University of Southampton
Southampton, UK

More information about this series at
http://www.springer.com/series/14750

Valeriy Zakamulin

Market Timing with Moving Averages

The Anatomy and Performance of Trading Rules

Valeriy Zakamulin
School of Business and Law
University of Agder
Kristiansand
Norway

New Developments in Quantitative Trading and Investment
ISBN 978-3-319-60969-0   ISBN 978-3-319-60970-6   (eBook)
DOI 10.1007/978-3-319-60970-6

Library of Congress Control Number: 2017948682

Cover design by Samantha Johnson

Printed on acid-free paper

This Palgrave Macmillan imprint is published by Springer Nature
The registered company is Springer International Publishing AG
The registered company address is: Gewerbestrasse 11, 6330 Cham, Switzerland

I dedicate this book to all technical traders, especially newcomers to the market, and hope that it helps them better understand the tools at their disposal.

Preface

Motivation for Writing this Book

Over the course of the last decade, the author of this book has been interested in the stock return predictability which is one of the most controversial topics in financial research. The existence of stock return predictability is of great interest to both practitioners and academics alike. Traditionally, in finance literature the stock returns were predicted using various financial ratios and macroeconomic variables. Unfortunately, the evidence of stock return predictability by either financial ratios or macroeconomic variables is unconvincing. Technical analysis represents another methodology of predicting future stock returns through the study of past stock prices and uncovering some recurrent regularities, or patterns, in price dynamics.

Whereas technical analysis has been extensively used by traders for almost a century and the majority of active traders strongly believe in stock return predictability, academics had long been skeptical about the usefulness of technical analysis. Yet, the academics' attitude toward the technical analysis is gradually changing. The findings in a series of papers on technical analysis of financial markets suggest that one should not bluntly dismiss the value of technical analysis. Recently, we have witnessed a constantly increasing interest in technical analysis from both practitioners and academics alike. This interest developed because over the decade of 2000s, that covers two severe stock market downturns, many technical trading rules outperformed the market by a large margin.

One of the basic principles of technical analysis is that "prices move in trends." Traders firmly believe that these trends can be identified in a timely manner and used to generate profits and limit losses. Consequently, trend following is the most widespread trading strategy; it tries to jump on a trend and ride it. Specifically, when stock prices are trending upward (downward), it is time to buy (sell) the stock. The problem is that stock prices fluctuate wildly which makes it difficult for traders to identify the trend in stock prices. Moving averages are used to "smooth" the fluctuations in the stock price in order to highlight the underlying trend. As a matter of fact, a moving average is one of the oldest and most popular tools used in technical analysis for detecting a trend.

Over the course of the last few years, the author of this book has conducted research on the profitability of moving average trading rules. The outcome of this research was a collection of papers, two of which were published in scientific journals. The rest of the papers in this collection laid the foundations for this book on market timing with moving averages. In principle, there are already many books on technical analysis of financial markets that cover the subject of trading with moving averages. Why a new book on moving averages? The reasons for writing a new book are explained below.

All existing books on trading with moving averages can be divided into two broad categories:

1. Books that cover all existing methods, tools, and techniques used in technical analysis of financial markets (two examples of such books are Murphy 1999, and Kirkpatrick and Dahlquist 2010). In these books, that can be called as the "Bibles" of technical analysis, the topic on technical trading with moving averages is covered briefly and superficially; the authors give only the most essential information about moving averages and technical trading rules based on moving averages.
2. Books that are devoted solely to the subject of moving averages (examples of such books are Burns and Burns 2015, and Droke 2001). These books are usually written for beginners; the authors cover in all details only the most basic types of moving averages and technical trading rules based on moving averages.

Regardless of the book type, since the subject of technical trading with moving averages is constantly developing, the information in the existing books is usually outdated and/or obsolete. Thereby the existing books lack in-depth, comprehensive, and up-to-date information on technical trading with moving averages.

Unfortunately, the absence of a comprehensive handbook on technical trading with moving averages is just one of several issues with the subject. The other two important issues are as follows:

1. There are many types of moving averages as well as there are many technical trading rules based on one or several moving averages. As a result, technical traders are overwhelmed by the variety of choices between different types of trading rules and moving averages. One of the controversies about market timing with moving averages is over which trading rule in combination with which moving average(s) produces the best performance. The situation is further complicated because in order to compute a moving average one must specify the size of the averaging window. Again, there is a big controversy over the optimal size of this window for each trading rule, moving average, and financial market. The development in this field has consisted in proposing new ad hoc rules and using more elaborate types of moving averages in the existing rules without any deeper analysis of commonalities and differences between miscellaneous choices for trading rules and moving averages. It would be no exaggeration to say that the existing situation resembles total chaos and mess from the perspective of a newcomer to this field.

2. Virtually, all existing books and the majority of papers on technical trading with moving averages claim that one can easily beat the market and become rich by using moving averages. For example, in one popular paper the author claims that using moving averages in the stock market produces "equity-like returns with bond-like volatility and drawdowns" (i.e., moving averages produce stock-like returns with bond-like risk). There are many similar claims about the allegedly superior performance of moving average trading strategies. The major problem is that all these claims are usually supported by colorful narratives and anecdotal evidence rather than objective scientific evidence. At best, such claims are "supported" by performing a simple back-test using an arbitrary and short historical sample of data and reporting the highest observed performance of a trading rule. Yet, serious researchers know very well that the observed performance of the best trading rule in a back-test severely overestimates its real-life performance.

Overall, despite a series of publications in academic journals, modern technical analysis in general and trading with moving averages in particular still remain art rather than science. In the absence of in-depth analysis of commonalities and differences between various trading rules and moving averages, technical traders do not really understand the response characteristics of the trading indicators they use and the selection of a specific trading rule, coupled with some specific type of a moving average, is made based mainly on intuition and anecdotal evidence. Besides, there is usually no objective scientific evidence which supports the claim that some specific moving average trading strategy allows one to beat the market.

To the best knowledge of the author, there is only one book to date (Aronson 2010) that conveys the idea that all claims in technical analysis represent, in principle, scientific testable claims. The book describes carefully all common pitfalls in back-testing trading rules and presents correct scientific methods of testing the profitability of technical trading rules. The book contains a thorough review of statistical principles with a brief case study of profitability of various technical trading rules (including a few moving average trading rules) in one specific financial market. Therefore, whereas the book makes a very good job in explaining how to scientifically evaluate the performance of trading rules, the case study in the book is very limited; the question of how profitable are the moving average trading rules in various financial markets remains unanswered.

Book Objectives and Structure

Given the increasing popularity of trading with moving averages, we thought of writing this book in order to overcome the shortcomings of existing books and give the readers the most comprehensive and objective information about this topic. Specifically, the goals of this book are threefold:

1. Provide the in-depth coverage of various types of moving averages, their properties, and technical trading rules based on moving averages.
2. Uncover the anatomy of market timing rules with moving averages and offer a new and very insightful reinterpretation of the existing rules.
3. Revisit the myths regarding the superior performance of moving average trading rules and provide the reader with the most objective assessment of the profitability of these rules in different financial markets.

This book is composed of four parts and a concluding chapter; each part consists of two or three chapters:

Part I: This part provides the in-depth coverage of various types of moving averages and their properties.

Chapter 1: This chapter presents a brief motivation for using moving averages for trend detection, how moving averages are computed, and their two key properties: the average lag (delay) time and smoothness. The most important thing to understand right from the start is that there is a direct relationship between the average lag time and smoothness of a moving average.

Chapter 2: This chapter introduces the notion of a general weighted moving average and shows that each specific moving average can be uniquely characterized by either a price weighting function or a price-change weighting function. It also demonstrates how to quantitatively assess the average lag time and smoothness of a moving average. Finally, the analysis provided in this chapter reveals two important properties of moving averages when prices trend steadily.

Chapter 3: This chapter presents a detailed review of all ordinary types of moving averages, as well as some exotic types of moving averages. These exotic moving averages include moving averages of moving averages and mixed moving averages with less average lag time. For the majority of moving averages, this chapter computes the closed-form solutions for the average lag time and smoothness. This chapter also demonstrates that the average lag time of a moving average can easily be manipulated; therefore, the notion of the average lag time has very little to do with the delay time in the identification of turning points in a price trend.

Part II: This part reviews the technical trading rules based on moving averages and uncovers the anatomy of these rules.

Chapter 4: This chapter reviews the most common trend-following rules that are based on moving averages of prices. It also discusses the principles behind the generation of trading signals in these rules. This chapter also illustrates the limitations of these rules and argues that the moving average trading rules are advantageous only when the trend is strong and long lasting.

Chapter 5: This key chapter presents a methodology for examining how the trading signal in a moving average rule is computed. Then using this methodology, the chapter examines the computation of trading signals in all moving average rules and investigates the commonalities and differences between the rules. The main conclusion that can be drawn from this study is that the computation of the trading indicator in every rule, based on either one or multiple moving averages, can equivalently be interpreted as the computation of a single weighted moving average of price changes. The analysis presented in this chapter uncovers the anatomy of moving average trading rules, provides very useful insights about popular trend rules, and offers a new reinterpretation of the existing moving average trading rules.

Part III: In this part, we present our methodology for how to scientifically test the claim that one can beat the market by using moving average trading rules.

Chapter 6: This chapter starts with a review of transaction costs in capital markets. Then it demonstrates how to simulate the returns to a moving average trading strategy in the presence of transaction costs. The following two cases are considered when a trading indicator generates a sell signal: case one where the trader switches to cash, and case two where the trader alternatively sells short a financial asset.

Chapter 7: This chapter explains how to evaluate the performance of a trading strategy and how to carry out a statistical test of the hypothesis that a moving average trading strategy outperforms the corresponding buy-and-hold strategy. In particular, it argues that there is no unique performance measure, reviews the most popular performance measures, and points to the limitations of these measures. The chapter then surveys the parametric methods of testing the outperformance hypothesis and the current "state of the art" non-parametric methods.

Chapter 8: Technical traders typically rely on back-testing which is defined as the process of testing a trading strategy using relevant historical data. Back-testing usually involves "data mining" which denotes the practice of finding a profitable trading strategy by extensive search through a vast number of alternative strategies. This chapter explains that the data-mining procedure tends to find a strategy which performance benefited most from luck. As a result, the performance of the best strategy in a back-test is upward biased. This fact motivates that any back-test must be combined with a data-mining correction procedure that adjusts downward the estimated performance. Another straightforward method of the estimation of true performance of a trading strategy is to employ a validation procedure; this method is called forward-testing.

Part IV: This part contains case studies of profitability of moving average trading rules in different financial markets.

Chapter 9: This chapter utilizes the longest historical sample of data on the S&P Composite stock index and comprehensively evaluates the profitability of various moving average trading rules. Among other things, the chapter investigates the following: which trading rules performed best; whether the choice of moving average influences the performance of trading rules; how accurately the trading rules identify the bullish and bearish stock market trends; whether there is any advantage in trading daily rather than monthly; and how persistent is the outperformance delivered by the moving average trading rules. The results of this study allow us to revisit the myths regarding the superior performance of the moving average trading rules in this well-known stock market and fully understand their advantages and disadvantages.

Chapter 10: This chapter tests the profitability of various moving average trading rules in different financial markets: stocks, bonds, currencies, and commodities. The results of these tests allow us to better understand the properties of the moving average trading strategies and find out which trading rules are profitable in which markets. The chapter concludes with a few practical recommendations for traders testing the profitability of moving average trading rules. The analysis presented in this chapter also suggests a hypothesis about simultaneous existence, in the same financial market, of several trends with different durations.

Conclusion, Chapter 11: This concluding chapter presents a brief summary of the key contributions of this book to the field of technical analysis of financial markets. In addition, the chapter derives an alternative representation of the main result on the anatomy of moving average trading rules. It is demonstrated that all these rules predict the future price trend using a simple linear forecasting model that is identical to models used in modern empirical finance. Therefore, this alternative representation allows us to reconcile modern empirical finance with technical analysis of financial markets that uses moving averages. Finally, this chapter discusses whether the advantages of the moving average rules, observed using past (historical) data, are likely to persist in the future.

Readership and Prerequisites

This book is not for a layman who believes that moving averages offer a simple, quick, and easy way to riches. This book is primarily intended for a serious and mathematically minded reader who wants to get an in-depth knowledge of the subject. Even though, for the sake of completeness of exposition, we briefly cover all relevant theoretical topics, we do not explain the basic financial terminology, notions, and jargons. Therefore, this book is best suited for the reader with an MS degree in economics or business administration who is familiar with basic concepts in investments and statistics. Examples of such readers are academics, students at economic departments, and practitioners (portfolio managers, quants, traders, etc.). This book is, in principle, also suited for self-study by strongly motivated readers without prior exposure to finance theory, but in this case the book should be supplemented by an introductory textbook on investments at least (an example of such book is Bodie, Kane, and Marcus 2007).

Parts I and II are relatively easy to comprehend. These parts require only the knowledge of high school mathematics, basically a familiarity with arithmetic and geometric series and their sums. The material presented in Parts III and IV of this book makes it necessary to use extensively financial mathematics and statistics. Without the required prerequisites, the reader can try to skip Part III of the book and jump directly to Part IV. However, in order to understand the results reported in Part IV of this book, the reader is required to have a superficial knowledge of back-tests and forward-tests, and to understand our notion of "outperformance" which is the difference between the performances of the moving average trading strategy and the corresponding buy-and-hold strategy.

Supplementary Book Materials

The author of this book provides two types of supplementary book materials that are available online on the author's website http://vzakamulin. weebly.com/.

The first type of supplementary book materials is interactive Web applications. Interactivity means that outputs in these applications change instantly as the user modifies the inputs. Therefore, these applications not only replicate the illustrations and results provided in this book, but also allow the user to modify inputs and get new illustrations and results. Last but not least, these applications offer the user real-time trading signals for some stock market indices. There are no prerequisites for using the first type of supplementary book materials.

The results reported in this book were obtained using the open source programming language **R** (see https://www.r-project.org/). To let anyone reproduce some of the results provided in this book, as well as test the profitability of moving average trading rules using own data, the author provides the second type of supplementary book materials: two R packages that include reusable R functions, the documentation that describes how to use them, and sample data. The first R package is bbdetection that allows the user to detect bull and bear states in a financial market and to get the dating and the descriptive statistics of these states. The second R package is matiming that allows the user to simulate the returns to different moving average trading rules and to perform both back-tests and forward-tests of the trading rules. The prerequisites for using the second type of supplementary book materials are the familiarity with R language and the ability to write R programs.

Kristiansand, Norway   Valeriy Zakamulin

References

Aronson, D. (2010). *Evidence-based technical analysis: Applying the scientific method and statistical inference to trading signals.* John Wiley & Sons, Ltd.

Bodie, Z., Kane, A., & Marcus, A. J. (2007). Investments. *McGraw Hill.*

Burns, S. & Burns, H. (2015). *Moving averages 101: Incredible signals that will make you money in the stock market.* CreateSpace Independent Publishing Platform.

Droke, C. (2001). *Moving Averages Simplified,* Marketplace Books.

Kirkpatrick, C. D. & Dahlquist, J. (2010). *Technical analysis: The complete resource for financial market technicians.* FT Press.

Murphy, J. J. (1999). *Technical analysis of the financial markets: A comprehensive guide to trading methods and applications.* New York Institute of Finance.

Acknowledgements

This book is based on the research conducted by the author over the course of several years. As with any book, this book is the product not only of its author, but also of his colleagues, his environment, of the encouragements, support, and discussions with different people who, voluntarily or not, contributed to this book. While it is enormously difficult to do justice to all relevant persons, the author would like to thank explicitly the following individuals and groups.

The key result on the anatomy of moving average trading rules appeared due to the author's discussions with Henry Stern back in 2013. These discussions stimulated the author to think deeply on the differences and similarities between various technical trading rules based on moving averages of prices. Over the years, the author has greatly profited from inspiring discussions and collaborations with Steen Koekebakker. The author is indebted to Michael Harris (of Price Action Lab) for his constructive feedback on the first draft of this book. Comments and encouragements from Wesley Gray (of Alpha Architect) are greatly acknowledged. The author is also grateful to helpful comments, discussions, and suggestions from the participants at the conferences where the author presented his papers on market timing with moving averages.

All the empirical studies in the book were conducted using the R programming language. The author expresses a thought of gratitude to the R Development Core Team for creating powerful and free statistical software and to the RStudio developers for their excellent integrated development environment for R. The author supplies with this book two R packages and thanks a group of master students, whom the author supervised during the spring of 2017, for testing these packages. Special thanks to Aimee Dibbens,

Tula Weis, Nicole Tovstiga, and their colleagues at Palgrave for welcoming this book's proposals and their excellent help and professionalism in dealing with the book publication issues. Last but not least, the author would like to thank his family for their love and support through the years.

Contents

Part I   Moving Averages

1   Why Moving Averages?   3

2   Basics of Moving Averages   11

3   Types of Moving Averages   23

Part II   Trading Rules and Their Anatomy

4   Technical Trading Rules   55

5   Anatomy of Trading Rules   71

Part III   Performance Testing Methodology

6   Transaction Costs and Returns to a Trading Strategy   105

7   Performance Measurement and Outperformance Tests   111

8   Testing Profitability of Technical Trading Rules   129

Part IV   Case Studies

  9   Trading the Standard and Poor's Composite Index   143

10   Trading in Other Financial Markets   223

11   Conclusion   265

Index   275

About the Author

Valeriy Zakamulin is Professor of Finance at the School of Business and Law, University of Agder, Norway, where he teaches graduate courses in Finance. His first graduate academic degree is an MS in Radio Engineering. After receiving this degree, Valeriy Zakamulin had been working for many years as a research fellow at a computer science department, developing both computer hardware and software. Later on, Valeriy Zakamulin received an MS in Economics and Business Administration and a Ph.D. in Finance. He has published more than 30 articles in various refereed academic and practitioner journals and is a frequent speaker at international conferences. He has also served on editorial boards of several economics and finance journals. His current research interests cover behavioral finance, portfolio optimization, time-series analysis of financial data, financial asset return and risk predictability, and technical analysis of financial markets.

List of Figures

Fig. 1.1   Noisy price is smoothed by a centered moving average   6

Fig. 1.2   Noisy price is smoothed by a right-aligned moving average   8

Fig. 2.1   Illustration of the lag time between the time series of stock
            prices and the moving average of prices   17

Fig. 3.1   LMA and SMA applied to the monthly closing prices of the
            S&P 500 index   27

Fig. 3.2   Weighting functions of LMA and SMA with the same lag time
            of 5 periods. In the price weighting functions, Lag i denotes
            the lag of P_{t-i}. In the price-change weighting functions, Lag j
            denotes the lag of ΔP_{t-j} 28

Fig. 3.3   Illustration of the behavior of $SMA(11)$ and $LMA(16)$ along
            the stock price trend and their reactions to a sharp change
            in the trend. Note that when prices trend upward or
            downward, the values of the two moving averages coincide.
            The differences between the values of these two moving
            averages appear during the period of their "adaptation" to the
            change in the trend   29

Fig. 3.4   Weighting functions of EMA and SMA with the same lag time
            of 5 periods. In the price weighting functions, Lag i denotes
            the lag of P_{t-i}. In the price-change weighting functions, Lag j
            denotes the lag of ΔP_{t-j}. The weights of the (infinite) EMA are
            cut off at lag 21   33

Fig. 3.5   EMA and SMA applied to the monthly closing prices of the
            S&P 500 index   34

Fig. 3.6   Illustration of the behavior of $SMA(11)$ and $EMA(11)$ along
            the stock price trend and their reactions to a sharp change in
            the trend   35

Fig. 3.7   Price weighting functions of $SMA_{11}(P)$ and $SMA_{11}(SMA_{11}(P))$ 36

Fig. 3.8 $TMA_{11}(P)$ and $SMA_{11}(P)$ applied to the monthly closing prices of the S&P 500 index   37

Fig. 3.9   Price weighting functions of $EMA_{11}(P), EMA_{11}(EMA_{11}(P))$, and $EMA_{11}(EMA_{11}(EMA_{11}(P)))$. The weights of the (infinite) EMAs are cut off at lag 30   38

Fig. 3.10   Price weighting functions of EMA_{11} and ZLEMA based on EMA_{11}. The weights of the (infinite) EMAs are cut off at lag 21   41

Fig. 3.11 EMA_{11} and ZLEMA based on EMA_{11} applied to the monthly closing prices of the S&P 500 index   41

Fig. 3.12   Illustration of the behavior of EMA_{11} and ZLEMA (based on EMA_{22}) along the stock price trend and their reactions to a sharp change in the trend. Both EMA_{11} and ZLEMA have the same lag time of 3 periods in the detection of the turning point in the trend   42

Fig. 3.13   Price weighting functions of EMA_{11}, DEMA based on EMA_{11}, and TEMA based on EMA_{11}. The weights of the (infinite) EMAs are cut off at lag 21   44

Fig. 3.14 EMA_{11}, DEMA and TEMA (both of them are based on EMA_{11}) applied to the monthly closing prices of the S&P 500 index   44

Fig. 3.15   Weighting functions of LMA_{16} and HMA based on $n = 16$ 46

Fig. 3.16 LMA_{16} and HMA (based on $n = 16$) applied to the monthly closing prices of the S&P 500 index   46

Fig. 4.1   Trading with 200-day Momentum rule. The *top panel* plots the values of the S&P 500 index over the period from January 1997 to December 2006. The *shaded areas* in this plot indicate the periods where this rule generates a Sell signal. The *bottom panel* plots the values of the technical trading indicator   57

Fig. 4.2   Trading based on the change in 200-day EMA. The *top panel* plots the values of the S&P 500 index over the period from January 1997 to December 2006, as well as the values of EMA (200). The *shaded areas* in this plot indicate the periods where this rule generates a Sell signal. The *bottom panel* plots the values of the technical trading indicator   58

Fig. 4.3   Trading with 200-day Simple Moving Average. The *top panel* plots the values of the S&P 500 index over the period from January 1997 to December 2006, as well as the values of SMA (200). The *shaded areas* in this plot indicate the periods where this rule generates a Sell signal. The *bottom panel* plots the values of the technical trading indicator   59

Fig. 4.4   Trading with 50/200-day Moving Average Crossover. The *top panel* plots the values of the S&P 500 index over the period from January 1997 to December 2006, as well as the values of

SMA(50) and SMA(200). The *shaded areas* in this plot indicate
the periods where this rule generates a Sell signal. The *bottom
panel* plots the values of the technical trading indicator   61

Fig. 4.5   Illustration of a moving average ribbon as well as the common
interpretation of the dynamics of multiple moving averages in a
ribbon   62

Fig. 4.6   Trading with 12/29/9-day Moving Average
Convergence/Divergence rule. The *top panel* plots the values
of the S&P 500 index and the values of 12- and 29-day EMAs.
The *shaded areas* in this panel indicate the periods where this
rule generates a Sell signal. The *middle panel* plots the values of
MAC(12,29) and EMA(9,MAC(12,29)). The *bottom panel*
plots the values of the technical trading indicator of the
MACD(12,29,9) rule   64

Fig. 4.7   Trading with 200-day Simple Moving Average. The figure
plots the values of the S&P 500 index over the period from
July 1999 to October 2000, as well as the values of SMA(200).
The shaded areas in this plot indicate the periods where this
rule generates a Sell signal   66

Fig. 4.8   Trading with 50/200-day Moving Average Crossover. The
figure plots the values of the S&P 500 index over the period
from January 1998 to December 1998, as well as the values of
50- and 200-day SMAs. The *shaded area* in this plot indicates
the period where this rule generates a Sell signal   67

Fig. 5.1   The shapes of the price change weighting functions in the
Momentum (MOM) rule and four Price Minus Moving
Average rules: Price Minus Simple Moving Average (P-SMA)
rule, Price Minus Linear Moving Average (P-LMA) rule, Price
Minus Exponential Moving Average (P-EMA) rule, and Price
Minus Triangular Moving Average (P-TMA) rule. In all rules,
the size of the averaging window equals $n = 30$. The weights
of the price changes in the P-EMA rule are cut off at lag 30   79

Fig. 5.2   The shapes of the price change weighting functions in five
Moving Average Change of Direction rules: Simple Moving
Average (SMA) Change of Direction rule, Linear
(LMA) Moving Average Change of Direction rule, Exponential
Moving Average (EMA) Change of Direction rule, Double
Exponential Moving Average (EMA(EMA)) Change of
Direction rule, and Triangular (TMA) Moving Average
Change of Direction rule. In all rules, the size of the averaging
window equals $n = 30$. The weights of the price changes in the
ΔEMA and ΔEMA(EMA) rules are cut off at lag 30   82

Fig. 5.3   The shapes of the price change weighting functions in five
           Moving Average Crossover rules: Simple Moving Average
           (SMA) Crossover rule, Linear (LMA) Moving Average
           Crossover rule, Exponential Moving Average (EMA) Crossover
           rule, Double Exponential Moving Average (EMA(EMA))
           Crossover rule, and Triangular (TMA) Moving Average
           Crossover rule. In all rules, the sizes of the shorter and longer
           averaging windows equal $s = 10$ and $l = 30$ respectively   85
Fig. 5.4   The shapes of the price change weighting functions in five
           Simple Moving Average Crossover (SMAC) rules. In all rules,
           the size of the longer averaging window equals $l = 30$, whereas
           the size of the shorter averaging window takes values in
           $s \in [1, 5, 15, 25, 29]$ 87
Fig. 5.5   The shapes of the price change weighting functions for the
           Moving Average Crossover rule based on the Double
           Exponential Moving Average (DEMA) and the Triple
           Exponential Moving Average (TEMA) proposed by Patrick
           Mulloy, the Hull Moving Average (HMA) proposed by Alan
           Hull, and the Zero Lag Exponential Moving Average
           (ZLEMA) proposed by Ehlers and Way. In all rules, the sizes
           of the shorter and longer averaging windows equal $s = 10$ and
           $l = 30$ respectively   88
Fig. 5.6   The shape of the price change weighting functions in three
           Moving Average Convergence/Divergence rules: the original
           MACD rule of Gerald Appel based on using Exponential
           Moving Averages (EMA), and two MACD rules of Patrick
           Mulloy based on using Double Exponential Moving Averages
           (DEMA) and Triple Exponential Moving Averages (TEMA).
           In all rules, the sizes of the averaging windows equal $s = 12$,
           $l = 26$, and $n = 9$ respectively   90
Fig. 5.7   The shape of the price change weighting functions in three
           $MA_s(P - MA_n)$ rules. In all rules, the sizes of the shorter and
           longer averaging windows equal $s = 10$ and $n = 26$ respectively   93
Fig. 5.8   The shape of the price change weighting functions in three
           $MA_s(\Delta MA_n)$ rules. In all rules, the sizes of the shorter and
           longer averaging windows equal $s = 10$ and $n = 30$ respectively   94
Fig. 7.1   The standard deviation - mean return space and the capital
           allocation lines (CALs) through the risk-free asset r and two
           risky assets A and B 116
Fig. 8.1   Illustration of the out-of-sample testing procedure with an
           expanding in-sample window (*left panel*) and a rolling
           in-sample window (*right panel*). OOS denotes the
           out-of-sample segment of data for each in-sample segment   135

Fig. 9.1   The log of the S&P Composite index over 1857–2015 (*gray line*) versus the fitted segmented model (*black line*) given by $\log(I_t) = \log(I_0) + \mu t + \delta(t - t^*)^+ + \varepsilon_t$, where t^* is the breakpoint date, μ is the growth rate before the breakpoint, and $\mu + \delta$ is the growth rate after the breakpoint. The estimated breakpoint date is September 1944   147

Fig. 9.2   Bull and bear markets over the two historical sub-periods: 1857–1943 and 1944–2015. *Shaded areas* indicate bear market phases   153

Fig. 9.3   The shapes of the price-change weighting functions of the best trading strategies in a back test   167

Fig. 9.4   Rolling 10-year outperformance produced by the best trading strategies in a back test over the total historical period from January 1860 to December 2015. The first point in the graph gives the outperformance over the first 10-year period from January 1860 to December 1869. Outperformance is measured by $\Delta = SR_{MA} - SR_{BH}$ where SR_{MA} and SR_{BH} are the Sharpe ratios of the moving average strategy and the buy-and-hold strategy respectively   168

Fig. 9.5   The top 20 most frequent trading rules in a rolling back test. A 10-year rolling window is used to select the best performing strategies over the full sample period from 1860 to 2015   170

Fig. 9.6   Cluster dendrogram that shows the relationship between the 20 most frequent trading strategies in a rolling back test   171

Fig. 9.7   Rolling 10-year out-of-sample outperformance produced by the trading strategies simulated using both a rolling and an expanding in-sample window. The out-of-sample segment cover the period from January 1870 to December 2015. The first point in the graph gives the outperformance over the first 10-year period from January 1870 to December 1879. Outperformance is measured by $\Delta = SR_{MA} - SR_{BH}$ where SR_{MA} and SR_{BH} are the Sharpe ratios of the moving average strategy and the buy-and-hold strategy respectively   175

Fig. 9.8   *Upper panel* plots the out-of-sample outperformance of the moving average trading strategy for different choices of the sample split point. The outperformance is measured over the period that starts from the observation next to the split point and lasts to the end of the sample in December 2015. The *lower panel* of this figure plots the p-value of the test for outperformance. In particular, the following null hypothesis is tested: $H_0 : SR_{MA} - SR_{BH} \leq 0$ where SR_{MA} and SR_{BH} are the Sharpe ratios of the moving average strategy and the buy-and-hold strategy respectively. The *dashed horizontal line* in the *lower panel* depicts the location of the 10% significance level   177

Fig. 9.9   *Upper panel* plots the out-of-sample outperformance of the moving average trading strategy for different choices of the sample start point. Regardless of the sample start point, the out-of-sample segment covers the period from January 2000 to December 2015. The *lower panel* of this figure plots the p-value of the test for outperformance. In particular, the following null hypothesis is tested: $H_0 : SR_{MA} - SR_{BH} \leq 0$ where SR_{MA} and SR_{BH} are the Sharpe ratios of the moving average strategy and the buy-and-hold strategy respectively. The *dashed horizontal line* in the *lower panel* depicts the location of the 10% significance level   178

Fig. 9.10   *Upper panel* plots the cumulative returns to the P-SMA strategy versus the cumulative returns to the buy-and-hold strategy (B&H) over the out-of-sample period from January 1944 to December 2015. *Lower panel* plots the drawdowns to the P-SMA strategy versus the drawdowns to the buy-and-hold strategy over the out-of-sample period   183

Fig. 9.11   Mean returns and standard deviations of the buy-and-hold strategy and the moving average trading strategy over bull and bear markets. **BH** and **MA** denote the buy-and-hold strategy and the moving average trading strategy respectively   184

Fig. 9.12   Bull and Bear markets versus Buy and Sell signals generated by the moving average trading strategy. *Shaded ares* in the *upper part* of the plot indicate Sell periods. *Shaded areas* in the *lower part* of the plot indicate Bear market states   186

Fig. 9.13   Rolling 10-year outperformance produced by the SMAE (200,3.75) strategy and the SMAC(50,200) strategy over period from January 1930 to December 2015. The first point in the graph gives the outperformance over the first 10-year period from January 1930 to December 1939. Outperformance is measured by $\Delta = SR_{MA} - SR_{BH}$ where SR_{MA} and SR_{BH} are the Sharpe ratios of the moving average strategy and the buy-and-hold strategy respectively   191

Fig. 9.14   Rolling 10-year outperformance produced by the MOM(2) strategy in the absence of transaction costs over the period from January 1927 to December 2015. Outperformance is measured by $\Delta = SR_{MA} - SR_{BH}$ where SR_{MA} and SR_{BH} are the Sharpe ratios of the moving average strategy and the buy-and-hold strategy respectively   192

Fig. 9.15   Rolling 10-year outperformance produced by the EMACD
             (12,29,9) strategy over the period from January 1930 to
             December 2015. The first point in the graph gives the
             outperformance over the first 10-year period from January
             1930 to December 1939. Outperformance is measured by $\Delta =
             SR_{MA} - SR_{BH}$ where SR_{MA} and SR_{BH} are the Sharpe ratios
             of the moving average strategy and the buy-and-hold strategy
             respectively   195
Fig. 9.16   Empirical probability distribution functions of 2-year returns
             on the buy-and-hold strategy and the moving average strategy.
             BH denotes the buy-and-hold strategy, whereas **MA** denotes
             the moving average trading strategy   207
Fig. 9.17   Cumulative returns to the moving average strategy versus
             cumulative returns to the 60/40 portfolio of stocks and bonds
             over January 1944 to December 2011. **MA** denotes the
             moving average strategy whereas **60/40** denotes the 60/40
             portfolio of stocks and bonds. The returns to the moving
             average strategy are simulated out-of-sample using an
             expanding in-sample window. The initial in-sample period is
             from January 1929 to December 1943   210
Fig. 10.1   Rolling 10-year outperformance produced by the MOM(2)
             strategy over the period from January 1927 to December 2015.
             The first point in the graph gives the outperformance over the
             first 10-year period from January 1927 to December 1936.
             The returns to the MOM(2) strategy are simulated assuming
             daily trading without transaction costs. Outperformance is
             measured by $\Delta = SR_{MA} - SR_{BH}$ where SR_{MA} and SR_{BH} are
             the Sharpe ratios of the moving average strategy and the
             buy-and-hold strategy respectively   228
Fig. 10.2   Rolling 10-year outperformance in daily trading small stocks
             produced by the moving average strategy simulated
             out-of-sample over the period from January 1944 to December
             2015. Outperformance is measured by $\Delta = SR_{MA} - SR_{BH}$
             where SR_{MA} and SR_{BH} are the Sharpe ratios of the moving
             average strategy and the buy-and-hold strategy respectively   232
Fig. 10.3   The *upper panel* plots the yield on the long-term US
             government bonds over the period from January 1926 to
             December 2011, whereas the *lower panel* plots the natural log
             of the long-term government bond index over the same period.
             Shaded areas in the *lower panel* indicate the bear market phases 234

Fig. 10.4   Rolling 10-year outperformance in trading the long-term
            bonds produced by the moving average strategy simulated
            out-of-sample over the period from January 1944 to December
            2011. Outperformance is measured by $\Delta = SR_{MA} - SR_{BH}$
            where SR_{MA} and SR_{BH} are the Sharpe ratios of the moving
            average strategy and the buy-and-hold strategy respectively   238
Fig. 10.5   A weighted average of the foreign exchange value of the U.S.
            dollar against a subset of the broad index currencies. *Shaded
            areas* indicate the bear market phases   242
Fig. 10.6   *Left panel* plots the bull and bear market cycles in the US/Japan
            exchange rate. *Right panel* plots the bull and bear market cycles
            in the US/South Africa exchange rate. *Shaded areas* indicate the
            bear market phases   246
Fig. 10.7   *Top panel* plots the bull-bear markets in the US/Sweden
            exchange rate over the period from 1984 to 2015. *Shaded areas*
            indicate the bear market phases. *Bottom panel* plots the 5-year
            rolling outperformance delivered by the combined moving
            average strategy   247
Fig. 10.8   *Top panel* plots the bull-bear cycles in the S&P 500 index over
            the period from 1971 to 2015. *Bottom panel* plots the bull-bear
            cycles in the Precious metals index over the same period.
            Shaded areas indicate the bear market phases   249
Fig. 10.9   *Top panel* plots the bull-bear markets in the Grains commodity
            index over the period from 1984 to 2015. *Bottom panel* plots
            the 5-year rolling outperformance delivered by the combined
            moving average strategy (where short sales are not allowed)   255
Fig. 10.10 Empirical first-order autocorrelation functions of k-day returns
            in the following financial markets: the US/UK exchange rate,
            the large cap stocks, and the small cap stocks   259

List of Tables

Table 5.1	Four main shapes of the price change weighting function in a trading rule based on moving averages of prices	96
Table 9.1	Descriptive statistics for the monthly returns on the S&P Composite index and the risk-free rate of return	145
Table 9.2	Bull and bear markets over the total sample period 1857–2015	152
Table 9.3	Descriptive statistics of bull and bear markets	154
Table 9.4	Rank correlations based on different performance measures	158
Table 9.5	The top 10 strategies according to each performance measure	159
Table 9.6	Comparative performance of trading strategies with and without short sales	161
Table 9.7	Comparative performance of trading rules with different types of moving averages	164
Table 9.8	Top 10 best trading strategies in a back test	166
Table 9.9	Descriptive statistics of the buy-and-hold strategy and the out-of-sample performance of the moving average trading strategy	174
Table 9.10	Descriptive statistics of the buy-and-hold strategy and the out-of-sample performance of the moving average trading strategies	181
Table 9.11	Descriptive statistics of the buy-and-hold strategy and the moving average trading strategy over bull and bear markets	184
Table 9.12	Top 10 best trading strategies in a back test over January 1944 to December 2015	190
Table 9.13	Descriptive statistics of the buy-and-hold strategy and the out-of-sample performance of the moving average trading strategies	193

Table 9.14   Probability of loss and mean return over different investment
horizons for three major asset classes   203

Table 9.15   Descriptive statistics of 2-year returns on several alternative
assets   209

Table 9.16   Results of the hypothesis tests on the stability of means
and standard deviations over two sub-periods of data   216

Table 9.17   Results of the estimations of the two alternative models using
the total sample period 1857–2015   218

Table 9.18   Estimated transition probabilities of the two-states Markov
switching model for the stock market returns over two
historical sub-periods: 1857–1943 and 1944–2015   220

Table 9.19   Results of the hypothesis testing on the stability of the
parameters of the two-states Markov switching model for
the stock market returns over the two sub-periods   221

Table 10.1   Top 10 best trading strategies in a back test   227

Table 10.2   Outperformance delivered by the moving average trading
strategies in out-of-sample tests   230

Table 10.3   Descriptive statistics of the buy-and-hold strategy and the
moving average trading strategy over bull and bear markets   231

Table 10.4   Top 10 best trading strategies in a back test   235

Table 10.5   Descriptive statistics of the buy-and-hold strategy and the
out-of-sample performance of the moving average trading
strategies   237

Table 10.6   Top 10 best trading strategies in a back test with monthly
trading   243

Table 10.7   Top 10 best trading strategies in a back test with daily
trading   244

Table 10.8   Outperformance delivered by the moving average trading
strategies in out-of-sample tests   245

Table 10.9   Descriptive statistics of the buy-and-hold strategy and the
out-of-sample performance of the moving average trading
strategies in trading the US/Sweden exchange rate   246

Table 10.10   List of commodity price indices and their components   251

Table 10.11   Top 10 best trading strategies in a back test   252

Table 10.12   Outperformance delivered by the moving average trading
strategies in out-of-sample tests   253

Table 10.13   Descriptive statistics of the buy-and-hold strategy and the
out-of-sample performance of the moving average trading
strategies in trading the Grains commodity index   254

Table 10.14   Descriptive statistics of 2-year returns on different financial
asset classes over 1986–2011   261

Part I

Moving Averages

1

Why Moving Averages?

1.1 Trend Detection by Moving Averages

There is only one way to make money in financial markets and this way is usually expressed by an often-quoted investment maxim "buy low and sell high". The implementation of this maxim requires determining the time when the price is low and the subsequent time when the price is high (or the reverse in case of shorting a financial asset). Traditionally, fundamental analysis and technical analysis are two methods of identifying the proper times for buying and selling stocks.

Fundamental analysis is based on the idea that at some times the price of a stock deviates from its true or "intrinsic" value. If the price of a stock is below (above) its intrinsic value, the stock is said to be "undervalued" ("overvalued") and it is time to buy (sell) the stock. Fundamental analysis uses publicly available information about the company "fundamentals" that can be found in past income statements and balance sheets issued by the company under investigation. By studying this information, analysts evaluate the future earnings and dividend prospects of the company as well as its risk. These estimates are used to assess the intrinsic value of the company. The intrinsic stock price can be calculated using the Dividend Discount Model (see, for example, Bodie et al. 2007, Chap. 18) or its modifications.

Technical analysis represents a methodology of forecasting the future price movements through the study of past price data and uncovering some recurrent regularities, or patterns, in price dynamics. One of the basic principles of technical analysis is that certain price patterns consistently reappear and tend to produce the same outcomes. Another basic principle of technical analysis says that "prices move in trends". Analysts firmly believe that these trends

© The Author(s) 2017
V. Zakamulin, *Market Timing with Moving Averages*, New Developments
in Quantitative Trading and Investment, DOI 10.1007/978-3-319-60970-6_1

can be identified in a timely manner and used to generate profits and limit losses. Consequently, trend following is the most widespread market timing strategy; it tries to jump on a trend and ride it. Specifically, when stock prices are trending upward (downward), it is time to buy (sell) the stock.

Even though trend following is very simple in concept, its practical realization is complicated. One of the major difficulties is that stock prices fluctuate wildly due to imbalances between supply and demand and due to constant arrival of new information about company fundamentals. These up-and-down fluctuations make it hard to identify turning points in a trend. Moving averages are used to "smooth" the stock price in order to highlight the underlying trend. This methodology of detecting the trend by filtering the noise comes from the time-series analysis where centered (or two-sided) moving averages are used. The same methodology is applied for predicting the future stock price movement. However, for the purpose of forecasting, right-aligned (a.k.a. one-sided or trailing) moving averages are used. These two types of moving averages are considered below.

1.2   Centered Moving Averages in Time-Series Analysis

It is relatively easy to detect a trend and identify the turnings points in a trend in retrospect, that is, looking back on past data. Denote by $\{P_1, P_2, \ldots, P_T\}$ a series of observations of the closing prices of a stock over some time interval. It is common to think about the time-series of P_t as comprising two components: a trend and an irregular component or "noise" (see, for example, Hyndman and Athanasopoulos 2013, Chap. 6). Then, if we assume an additive model, we can write

$$P_t = T_t + I_t, \tag{1.1}$$

where T_t is a trend and I_t is noise. The standard assumption is that noise represents short-term fluctuations around the trend. Therefore this noise can be removed by smoothing the data using a centered moving average.

Any moving average of prices is calculated using a fixed size data "window" that is rolled through time. The length of this window of data, also called the averaging period (or the lookback period in a trailing moving average), is the time interval over which the moving average is computed. Denote by n the size of the averaging window which consists of a center and two halves of size k such that $n = 2k + 1$. The computation of the value of a Centered Moving Average at time t is given by

$$MA_t^c(n) = \frac{P_{t-k} + \cdots + P_t + \cdots + P_{t+k}}{n} = \frac{1}{n}\sum_{i=-k}^{k} P_{t+i}. \qquad (1.2)$$

The value of the trend component is then the value of the centered moving average $T_t = MA_t^c(n)$.

The size of the averaging window n is selected to effectively remove the noise in the time-series. Consider two illustrative examples based on using artificial stock price data depicted in Fig. 1.1. In both examples, the stock price trend is given by two linear segments. First, the stock price trends upward, then downward. We add noise to the trend and this noise is given by a high frequency sine wave. As a result, we construct an artificial time-series of the stock price according to Eq. (1.1). Observe that the two components of the price series, T_t and I_t, are known. The goal of this illustration is to visualize the shape of a centered moving average and its location relative to the stock price trend.

The top panel in Fig. 1.1 depicts the noisy price, the (intrinsic) stock price trend, and the value of the centered moving average computed using a window of $n = 11$ price observations. Similarly, the bottom panel in Fig. 1.1 shows the same price and its intrinsic trend, but this time the centered moving average is computed using a window of 21 price observations. Notice that a window of 21 price observations effectively removes the noise in the data series. Even though the top in the shape of this moving average represents a smoothed version of the top in the shape of the intrinsic trend, the turning point in the trend can be easily determined. In contrast, the moving average with a window of 11 price observations retains some small fluctuations. As a result, in this case the turning point in the trend is still cumbersome to identify.

Our example reveals two basic properties of a centered moving average. First, the longer the size of the averaging window, the better a moving averages removes the noise in a data series and the easier it is to detect turning points in the trend. Second, regardless of the size of the averaging window, the shape of a centered moving average follows closely the underlying trend in a data series and turning points in a centered moving average coincide in time with turning points in the intrinsic trend.

1.3 Right-Aligned Moving Averages in Market Timing

Centered moving averages are used to detect a trend and identify turning points in a trend in past data. In market timing, on the other hand, analysts need to detect a trend and identify turning points in real time. Specifically, at current

Centered moving average of 11 prices

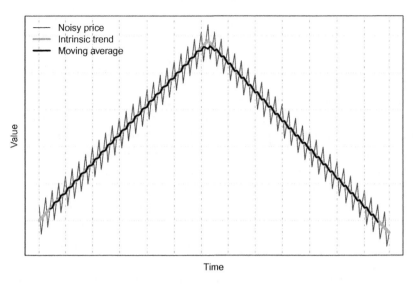

Centered moving average of 21 prices

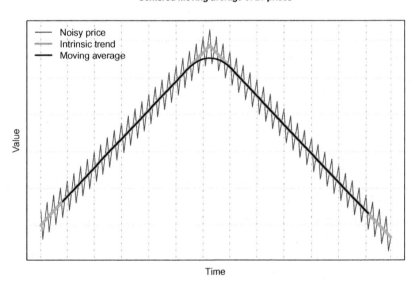

Fig. 1.1 Noisy price is smoothed by a centered moving average

time t analysts want to know the direction of the stock price trend. The new additional problem is that analysts know only the stock price data until t; the future stock prices from $t + 1$ and beyond are unknown. Therefore, at t one can use only the available data to compute the value of a moving average. In this case the value of a (Right-Aligned) Moving Average at t is computed as

$$MA_t^r(n) = \frac{P_t + P_{t-1} + \cdots + P_{t-n+1}}{n} = \frac{1}{n}\sum_{i=0}^{n-1}P_{t-i}. \qquad (1.3)$$

A comparison of the formulas for the calculation of the centered and right-aligned moving averages (given by Eqs. (1.2) and (1.3) respectively) reveals that the value of the right-aligned moving average at time t equals the value of the centered moving average at time $t - k$, where, recall, k denotes the half-size of the averaging window. Formally, this means the following identity:

$$MA_t^r(n) = MA_{t-k}^c(n).$$

Thus, a right-aligned moving average represents a lagged version of the centered moving average computed using the same size of the averaging window. Therefore a right-aligned moving average has the same smoothing properties as those of a centered moving average. Specifically, the longer the size of the averaging window in a right-aligned moving average, the better a moving average removes the noise in a data series. However, the longer the size of the averaging window, the longer the lag time. In particular, the lag time is given by

$$\text{Lag time } = k = \frac{n-1}{2}. \qquad (1.4)$$

These properties of a right-aligned moving average are illustrated in Fig. 1.2. This illustration uses the same artificial series of the stock price and the same sizes of the averaging window, 11 and 21 price observations, in the computation of the right-aligned moving average as in Fig. 1.1. Notice that in Fig. 1.2 the shapes of the moving averages of 11 and 21 prices are the same as in Fig. 1.1. Most importantly, observe that the more effective a right-aligned moving average smoothes the noise in the stock price data, the longer the lag time. The longer the lag time, the later a turning point in the stock price trend is detected by a moving average.

1.4 Chapter Summary

In the rest of the book, we consider only right-aligned moving averages that are used in timing a financial market. These averages are employed to detect the direction of the stock price trend and identify turning points in the trend in real time.

The profitability of a trend following strategy depends on the ability of early recognition of turning points in the stock price trend. However, since the stock

Right–aligned moving averages

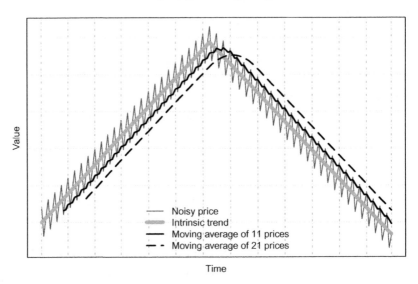

Fig. 1.2 Noisy price is smoothed by a right-aligned moving average

price is noisy, the noise complicates the identification of the trend and turning points in the trend. To remove the noise, analysts use trailing moving averages. These moving averages have the following two properties. First, the longer the size of the averaging window, the better a moving average removes the noise in the stock prices. At the same time, the longer the size of the averaging window, the longer the lag time between a turning point in the intrinsic stock price trend and the respective turning point in a moving average.

It is important to keep in mind that a turning point in a trend is identified with a delay. If analysts want to shorten the delay, they need to use a moving average with a shorter window size. Since moving averages with shorter windows remove the noise less effectively, using shorter windows leads to identification of many false turning points in the stock price trend. Increasing the size of the averaging window improves noise removal, but at the same time it also increases the delay time in recognizing the turning points in the stock price trend. Therefore the choice of the optimal size of the averaging window is crucial to the success of a trend following strategy. This choice needs to provide the optimal tradeoff between the lag time and the precision in the detection of true turnings points in a trend.

Last but not least, it is worth emphasizing that, since a trend is always recognized with some delay, the success of a trend following strategy also depends on the duration of a trend. That is, the duration of a trend should be long enough to make the trend following strategy profitable.

References

Bodie, Z., Kane, A., & Marcus, A. J. (2007). *Investments*. McGraw Hill.
Hyndman, R. J., & Athanasopoulos, G. (2013). *Forecasting: Principles and practice*. OTexts.

2

Basics of Moving Averages

In the preceding chapter we considered the simplest type of a moving average where equal weights are given to each price observation in the window of data. This chapter introduces the general weighted moving average and discusses how to quantitatively assess the two important characteristics of a moving average: the average lag time and the smoothness.

2.1 General Weighted Moving Average

Moving averages are computed using the averaging window of size n. Specifically, a moving average at time t is computed using the last closing price P_t and $n-1$ lagged prices P_{t-i}, $i \in [1, n-1]$. Generally, each price observation in the rolling window of data has its own weight in the computation of a moving average. More formally, a general weighted moving average of price series P at time t is computed as

$$MA_t(n, P) = \frac{w_0 P_t + w_1 P_{t-1} + w_2 P_{t-2} + \cdots + w_{n-1} P_{t-n+1}}{w_0 + w_1 + w_2 + \cdots + w_{n-1}} = \frac{\sum_{i=0}^{n-1} w_i P_{t-i}}{\sum_{i=0}^{n-1} w_i},$$

(2.1)

where w_i is the weight of price P_{t-i} in the computation of the weighted moving average. It is worth observing that, in order to compute a moving average, one has to use at least two prices; this means that one should have $n \geq 2$. Note that when the number of price observations used to compute a moving average equals one, a moving average becomes the last closing price, that is, $MA_t(1, P) = P_t$.

© The Author(s) 2017
V. Zakamulin, *Market Timing with Moving Averages*, New Developments in Quantitative Trading and Investment, DOI 10.1007/978-3-319-60970-6_2

The formula for a weighted moving average can alternatively be written as

$$MA_t(n, P) = \sum_{i=0}^{n-1} \psi_i P_{t-i}, \qquad (2.2)$$

where

$$\psi_i = \frac{w_i}{\sum_{j=0}^{n-1} w_j}.$$

Observe that weights ψ_i are normalized. Specifically, whereas the sum of weights w_i is not equal to one, it is easy to check that the sum of weights ψ_i equals one

$$\sum_{i=0}^{n-1} \psi_i = 1.$$

The set of weights given by either $\{w_0, w_1, \ldots, w_{n-1}\}$ or $\{\psi_0, \psi_1, \ldots, \psi_{n-1}\}$ is usually called a (price) "weighting function". Each type of a moving average has its own distinct weighting function. The most common shapes of a weighting function are: equal-weighting of prices, over-weighting the most recent prices, and hump-shaped form with under-weighting both the most recent and most distant prices.

The moving average is a linear operator. Specifically, if X and Y are two time series and a, b, and c are three arbitrary constants, then it is easy to prove the following property:

$$MA_t(n, aX + bY + c) = a \times MA_t(n, X) + b \times MA_t(n, Y) + c. \quad (2.3)$$

In the subsequent exposition, as a rule a moving average is computed using the series of prices P. Therefore, to shorten the notation, we will often drop the variable P in the notation of a moving average; that is, we will write $MA_t(n)$ instead of $MA_t(n, P)$.

2.2   Average Lag Time of a Moving Average

The weighting function of a moving average fully characterizes its properties and allows us to estimate the average lag time of the moving average. The idea behind the computation of the average lag time is to calculate the average "age" of the data included in the moving average.[1] In particular, the price observation at time $t - i$ has weight w_i in the calculation of a moving average and lags behind the most recent observation at time t by i periods. Consequently, the incremental delay from observation at $t - i$ amounts to $w_i \times i$. The average lag time is the lag time at which all the weights can be considered to be "concentrated". This idea yields the following identity:

$$\underbrace{(w_0 + w_1 + w_2 + \cdots + w_{n-1})}_{\text{Sum of all weights}} \times \text{Lag time}$$

$$= \underbrace{w_0 \times 0 + w_1 \times 1 + w_2 \times 2 + \cdots + w_{n-1} \times (n - 1)}_{\text{Weighted sum of delays of individual observations}}.$$

Therefore the average lag time of a weighted moving average can be computed using the following formula

$$\text{Lag time}(MA) = \frac{\sum_{i=1}^{n-1} w_i \times i}{\sum_{i=0}^{n-1} w_i} = \sum_{i=1}^{n-1} \psi_i \times i. \qquad (2.4)$$

Notice that since the most recent observation has the lag time 0, the weight w_0 disappears from the computation of the weighted sum of delays of individual observations.

The formula for the average lag time can be rewritten as follows. First, we write $\sum_{i=1}^{n-1} w_i \times i$ as a double sum (we just replace i with $\sum_{j=1}^{i} 1$)

$$\sum_{i=1}^{n-1} w_i \times i = \sum_{i=1}^{n-1} w_i \sum_{j=1}^{i} 1.$$

Second, interchanging the order of summation in the double sum above yields

$$\sum_{i=1}^{n-1} w_i \sum_{j=1}^{i} 1 = \sum_{j=1}^{n-1} \sum_{i=j}^{n-1} w_i.$$

---

[1]A similar idea is used in physics to compute the center of mass and in finance to compute the bond duration (Macaulay duration).

Finally, we rewrite the formula for the average lag time as

$$\text{Lag time}(MA) = \frac{\sum_{j=1}^{n-1} \sum_{i=j}^{n-1} w_i}{\sum_{i=0}^{n-1} w_i} = \sum_{j=1}^{n-1} \phi_j, \tag{2.5}$$

where the weight ϕ_j is given by

$$\phi_j = \frac{\sum_{i=j}^{n-1} w_i}{\sum_{i=0}^{n-1} w_i} = \sum_{i=j}^{n-1} \psi_j. \tag{2.6}$$

The usefulness of Eq. (2.5) will become clear shortly.

2.3   Alternative Representation of a Moving Average

The alternative representation of a moving average is motivated by the fact that a series of stock prices can be considered as a dynamic process in time. We introduce the notation

$$\Delta P_{t-i} = P_{t-i+1} - P_{t-i}$$

which is the change in the stock price over the time interval from $t - i$ to $t - i + 1$. Using this notation, we can write

$$P_{t-i} = P_t - \Delta P_{t-1} - \Delta P_{t-2} - \cdots - \Delta P_{t-i} = P_t - \sum_{j=1}^{i} \Delta P_{t-j}, \qquad i \geq 1.$$

The formula for the weighted moving average (given by Eq. (2.1)) can be rewritten as

$$MA_t(n) = \frac{w_0 P_t + \sum_{i=1}^{n-1} w_i \left(P_t - \sum_{j=1}^{i} \Delta P_{t-j} \right)}{\sum_{i=0}^{n-1} w_i}$$

$$= P_t - \frac{\sum_{i=1}^{n-1} w_i \sum_{j=1}^{i} \Delta P_{t-j}}{\sum_{i=0}^{n-1} w_i}.$$

Interchanging the order of summation in the double sum above yields

$$MA_t(n) = P_t - \frac{\sum_{j=1}^{n-1} \left(\sum_{i=j}^{n-1} w_i \right) \Delta P_{t-j}}{\sum_{i=0}^{n-1} w_i} = P_t - \sum_{j=1}^{n-1} \phi_j \Delta P_{t-j}, \quad (2.7)$$

where ϕ_j is given by Eq. (2.6). Therefore, all right-aligned moving averages can be represented as the last closing price minus the weighted sum of the previous price changes. Note that in the ordinary moving averages (to be considered in the next chapter) the weights are positive, $w_i > 0$ for all i. As a result, in this case the sequence of weights ϕ_j is decreasing with increasing j

$$\phi_1 > \phi_2 > \cdots > \phi_{n-1}.$$

Consequently, regardless of the shape of the weighting function for prices w_i, the weighting function ϕ_j always over-weights the most recent price changes.

In the subsequent exposition, we will call the weighting function ψ_i $(i \geq 0)$ the (normalized) "price weighting function" and the weighting function ϕ_j $(j \geq 1)$ the "price-change weighting function".

The alternative representation of a moving average provides very insightful information on the relationship between the stock price P_t, the value of the moving average $MA_t(n)$, and the average lag time. Therefore, let us elaborate more on this.

Equation (2.7) can be rewritten as

$$P_t - MA_t(n) = \sum_{j=1}^{n-1} \phi_j \Delta P_{t-j}.$$

This equation implies that the value of the moving average generally is not equal to the last closing price unless $\sum_{j=1}^{n-1} \phi_j \Delta P_{t-j} = 0$. For example, this happens when the price remains on the same level (the prices move sideways) in the averaging window. In this case $\Delta P_{t-j} = 0$ for all j and, as a result, the value of the moving average equals the last closing price.

If the prices move upward (downward) such that $\Delta P_{t-j} > 0$ $(\Delta P_{t-j} < 0)$ for all j, then $P_t - MA_t(n) > 0$ $(P_t - MA_t(n) < 0)$. *Therefore, when the prices are in uptrend, the moving average tends to be below the last closing price. In contrast, when the prices move downward, the moving average tends to be above the last closing price.* The stronger the trend, the larger the discrepancy between the last closing price and the value of a moving average.

Suppose that the change in the stock price follows a Random Walk process with a drift

$$\Delta P_{t-j} = E[\Delta P] + \sigma \varepsilon_j, \tag{2.8}$$

where $E[\Delta P]$ is the expected price change, σ is the standard deviation of the price change, and ε_j is a sequence of independent and identically distributed random variables with mean zero and unit variance ($E[\varepsilon_j] = 0$, $Var[\varepsilon_j] = 1$). In this case the expected difference between the last closing price and the value of the moving average equals

$$E\left[P_t - MA_t(n)\right] = E\left[\sum_{j=1}^{n-1} \phi_j \Delta P_{t-j}\right] = \sum_{j=1}^{n-1} \phi_j E[\Delta P_{t-j}]$$
$$= \text{Lag time}(MA) \times E[\Delta P], \tag{2.9}$$

where the last equality follows from Eq. (2.5). In words, the expected difference between the last closing price and the value of the moving average equals the average lag time times the average price change. Equation (2.9) is very insightful and implies that, in periods where variation in ΔP_{t-j} is rather small (for example, when prices are steadily increasing or decreasing), all moving averages with the same lag time move largely together *regardless of the shapes of their weighting functions and the sizes of their averaging windows.*[2] This property will be illustrated a number of times in the subsequent chapter.

It is instructive to illustrate graphically the relationship between the time series of stock prices, the moving average of prices, and the average lag time. For the sake of simplicity of illustration, we assume that the stock price steadily increases between times 0 and t. Specifically, we suppose that the stock price dynamic is given by

$$P_t = P_0 + \Delta P \times t,$$

where $\Delta P > 0$ is some arbitrary constant. The value of the moving average at time t is given by

$$MA_t(n) = P_t - \sum_{j=1}^{n-1} \phi_j \Delta P = P_0 + \Delta P \left(t - \sum_{j=1}^{n-1} \phi_j\right). \tag{2.10}$$

---

[2]Note that the average lag time is computed using the sequence of the weights ψ_i, $1 \leq i \leq n-1$. Many alternative sequences of weights can produce exactly the same value of the average lag time.

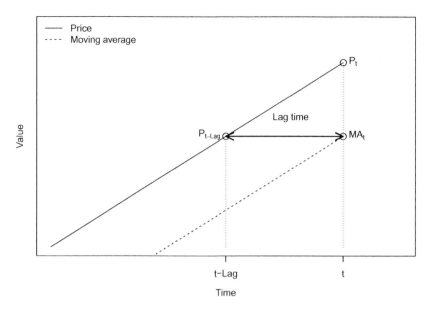

Fig. 2.1 Illustration of the lag time between the time series of stock prices and the moving average of prices

In this illustration, the lag time "Lag" between the time series of prices and the moving average of prices can be defined by the following relationship[3]

$$MA_t(n) = P_{t-\text{Lag}}.$$

This gives us the following equality

$$P_0 + \Delta P \left(t - \sum_{j=1}^{n-1} \phi_j \right) = P_0 + \Delta P(t - \text{Lag}).$$

The result is

$$\text{Lag} = \sum_{j=1}^{n-1} \phi_j,$$

which can be considered as an alternative derivation of the formula for the average lag time of a moving average. Graphically, the relationship between the stock price, the value of the moving average, and the average lag time is depicted in Fig. 2.1. It is important to emphasize that this relationship again implies

---

[3]In words, "Lag" is the required number of backshift operations applied to the time series of $\{MA_t(n)\}$ that makes it coincide with the time series of prices $\{P_t\}$.

that, when prices increase (or decrease) steadily, then all moving averages, that have exactly the same average lag time, move together regardless of the shapes of their weighting functions and the sizes of their averaging windows.

It is worth observing an additional interesting relationship between the dynamic of the price and the dynamic of a moving average of prices when prices increase or decrease steadily. Equation (2.10) implies that the change in the value of a moving average between times t and $t + 1$ is given by

$$\Delta MA_t(n) = MA_{t+1}(n) - MA_t(n) = \Delta P.$$

This is a very insightful result. In words, this result means that, when prices increase or decrease steadily (meaning that ΔP is virtually constant), the change in the value of a moving average equals the price change *regardless of the size of the averaging window and the shape of the weighting function.* That is, in this case both the price and all moving averages (with different average lag times) move parallel in a graph.

It is important to emphasize that the notion of the "average lag time" should be understood literally. That is, at each given moment the lag time depends on the weighting function of the moving average and the price changes in the averaging window. However, if we average over all specific lag times, then the average lag time will be given by Eq. (2.4) or alternatively by (2.5). Only in cases where the prices are steadily increasing or decreasing, the "average lag time" provides a correct numerical characterisation of the time lag between the price and the value of the moving average.

2.4   Smoothness of a Moving Average

Besides the average lag time, the other important characteristic of a moving average is its smoothness. The smoothness of a time series is often evaluated by analysing the properties of the first difference of the time series. In our context, to evaluate the smoothness of a moving average $MA_t(n)$, we start with the computation of the first difference

$$\Delta MA_t(n) = MA_{t+1}(n) - MA_t(n).$$

The idea is that the smoother the time series $MA_t(n)$ is, the lesser the variation in its first difference $\Delta MA_t(n)$. Using Eq. (2.2), the formula above can be rewritten as

$$\Delta MA_t(n) = \sum_{i=0}^{n-1} \psi_i P_{t+1-i} - \sum_{i=0}^{n-1} \psi_i P_{t-i} = \sum_{i=0}^{n-1} \psi_i \Delta P_{t-i}. \qquad (2.11)$$

One possible estimate of the smoothness of a moving average is the variance of $\Delta MA_t(n)$. In this case, small values of variance correspond to smoother series. If we assume that the change in the stock price follows a Random Walk process with a drift given by (2.8), then the variance of $\Delta MA_t(n)$ is equal to

$$\mathrm{Var}(\Delta MA_t(n)) = \sigma^2 \sum_{i=0}^{n-1} \psi_i^2 = \sigma^2 \times HI(MA), \qquad (2.12)$$

where

$$HI(MA) = \sum_{i=0}^{n-1} \psi_i^2$$

is the well-known Herfindahl index (a.k.a. Herfindahl-Hirschman Index, or HHI). This index is a commonly accepted measure of market concentration and competition among market participants. This index is also used to measure the investment portfolio concentration (see, for example, Ivkovic et al. 2008). Therefore Eq. (2.12) says that the variance of $\Delta MA_t(n)$ is directly proportional to the measure of concentration of weights in the price weighting function of a moving average and the variance of the price changes.[4]

The reciprocal of the Herfindahl index, $HI^{-1}(MA)$, computed using the (normalized) price weighting function of a moving average, represents a very convenient way to measure the smoothness of a moving average. The reasons for this are as follows. First, the properties of this index are well known. Second, to evaluate the smoothness, in this case one needs only to know the weighting function of a moving average; there is no need to estimate the smoothness empirically using some particular price series data. Third, in many cases it is possible to derive a closed-form solution for the smoothness of a specific moving average.

Using the properties of the Herfindahl index, the lowest smoothness of a moving average is attained when some $\psi_i = 1$ and all other weights are zero; in this case $HI = 1$. For some fixed n, the highest smoothness is attained when all weights are equal; in this case $HI = \frac{1}{n}$. That is, equal weighting of

---

[4]There is a large strand of econometric literature that demonstrates that volatility of financial assets is not constant over time. Specifically, there are alternating calm and turbulent periods in financial markets. Therefore, in real markets the smoothness of a moving average is not constant over time. In particular, the smoothness improves in calm periods and worsens in turbulent periods.

prices in a moving average produces the smoothest moving average for a given size n of the averaging window. As expected, when prices are equally weighted, increasing the size of the averaging window decreases the Herfindahl index and therefore increases the smoothness of a moving average.

2.5   Chapter Summary

Each specific moving average is uniquely characterized by its price weighting function. This price weighting function allows us to compute the two central characteristics of a moving average: the average lag time and smoothness. We demonstrated that the smoothing properties of a moving average can be evaluated by the inverse of the Herfindahl index. It turns out that both the average lag time and the Herfindahl index of a moving average are related to the concentration of weights in the price weighting function. Whereas the Herfindahl index directly measures the concentration of weights in the weighting function (the higher the concentration, the worse the smoothness), the average lag time provides the exact location of the weight concentration.

At each current time, the value of the moving average of prices generally deviates from the last closing price. Our analysis shows explicitly that when stock prices are steadily trending upward, the moving average lies below the price. In contrast, when stock prices are steadily trending downward, the moving average lies above the price.[5] On average, the discrepancy between the value of the moving average and the last closing price equals the average lag time times the average price change. Only when the prices are trending sideways (that is, they stay on about the same level) the value of the moving average is close to the last closing price.

The analysis provided in this chapter reveals two important properties of moving averages when prices trend steadily. The first property says that in this case all moving averages with the same average lag time move largely together (as a single moving average) regardless of the shapes of their weighting functions and the sizes of their averaging windows. As an immediate corollary to this property, the behavior of the moving averages with the same average lag time differs due to their different reactions to the changes in the stock price trend. The second property says that, when prices trend steadily, both the price and all moving averages (with different average lag times) move parallel in a graph regardless of the sizes of their averaging windows and the shapes of their weighting functions. As an immediate corollary to this property, a change in

---

[5]It is worth emphasizing that this relationship holds only when stock prices trend steadily in one direction. This relationship does not hold when the direction of the trend changes frequently.

the direction of the price trend causes moving averages with various average lag times to move in different directions in a graph.

Reference

Ivkovic, Z., Sialm, C., & Weisbenner, S. (2008). Portfolio concentration and the performance of individual investors. *Journal of Financial and Quantitative Analysis, 43*(3), 613–655.

3

Types of Moving Averages

In the preceding chapter we considered the general weighted moving average. This chapter aims to give an overview of some specific types of moving averages. However, since there is a huge amount of different types of moving averages, and this amount is constantly increasing, it is virtually impossible to review them all. Therefore, in this chapter we cover in all details the ordinary moving averages. In addition, we present some examples of exotic moving averages: moving averages of moving averages and mixed moving averages.

3.1 Ordinary Moving Averages

In this section we consider the most common types of moving averages used to time the market.

3.1.1 Simple Moving Average

The Simple Moving Average (SMA) computes the arithmetic mean of n prices. This type of moving average was considered in Chap. 1. For the sake of completeness of exposition, we repeat how this moving average is computed

$$SMA_t(n) = \frac{1}{n} \sum_{i=0}^{n-1} P_{t-i}. \tag{3.1}$$

© The Author(s) 2017
V. Zakamulin, *Market Timing with Moving Averages*, New Developments in Quantitative Trading and Investment, DOI 10.1007/978-3-319-60970-6_3

In this moving average, each price observation has the same weight $w_i = 1$ ($\psi_i = \frac{1}{n}$).

Note that the difference between the values of $SMA(n)$ at times t and $t-1$ equals

$$SMA_t(n) - SMA_{t-1}(n) = \frac{P_t - P_{t-n}}{n}.$$

Therefore the recursive formula for SMA is given by

$$SMA_t(n) = SMA_{t-1}(n) + \frac{P_t - P_{t-n}}{n}. \qquad (3.2)$$

This recursive formula can be used to accelerate the computation of SMA in practical applications. Specifically, the calculation of SMA according to formula (3.1) requires $n-1$ summations and one division; totally n operations. In contrast, the calculation of SMA according to formula (3.2) requires one summation, one subtraction, and one division; totally 3 operations regardless of the size of the averaging window.

The average lag time of SMA is given by (see the subsequent appendix for the details of the derivation)

$$\text{Lag time}(SMA_n) = \frac{\sum_{i=1}^{n-1} i}{\sum_{i=0}^{n-1} 1} = \frac{1 + 2 + \cdots + (n-1)}{n} = \frac{n-1}{2}. \qquad (3.3)$$

The Herfindahl index of SMA equals $\frac{1}{n}$; therefore the smoothness of $SMA(n)$, in our definition, equals $\left(\frac{1}{n}\right)^{-1} = n$. Obviously, increasing the size of the averaging window increases both the smoothness and the average lag time of SMA. The average lag time of SMA is a liner function of its smoothness

$$\text{Lag time}(SMA_n) = \frac{1}{2} \times \text{Smoothness}(SMA_n) - \frac{1}{2}. \qquad (3.4)$$

3.1.2 Linear Moving Average

The SMA is, in fact, an equally-weighted moving average where an equal weight is given to each price observation. Many analysts believe that the most recent stock prices contain more relevant information on the future direction of the stock price than earlier stock prices. Therefore, they argue, one should put more weight on the more recent price observations. Formally, this argument requires

$$w_0 > w_1 > w_2 > \cdots > w_{n-1}. \tag{3.5}$$

To correct the weighting problem in SMA, some analysts employ the linearly weighted moving average.

A Linear (or linearly-weighted) Moving Average (LMA) is computed as

$$LMA_t(n) = \frac{nP_t + (n-1)P_{t-1} + (n-2)P_{t-2}\ldots + P_{t-n+1}}{n + (n-1) + (n-2) + \cdots + 1} = \frac{\sum_{i=0}^{n-1}(n-i)P_{t-i}}{\sum_{i=0}^{n-1}(n-i)}. \tag{3.6}$$

In the linearly weighted moving average the weights decrease in arithmetic progression. In particular, in $LMA(n)$ the latest observation has weight n, the second latest $n-1$, etc. down to one.

The sum in the denominator of the fraction above equals

$$\Sigma = \sum_{i=0}^{n-1}(n-i) = \frac{n(n+1)}{2}.$$

The difference between the values of $LMA(n)$ at times t and $t-1$ equals

$$\Big(LMA_t(n) - LMA_{t-1}(n)\Big)\Sigma = nP_t - \underbrace{(P_{t-1} + P_{t-2} + \cdots + P_{t-n+1} + P_{t-n})}_{\text{Total}_{t-1}},$$

where Total_{t-1} denotes the time $t-1$ sum of the prices in the averaging window. Therefore the recursive computation of LMA is performed as follows. First, the new value of LMA is computed

$$LMA_t(n) = LMA_{t-1}(n) + \frac{nP_t - \text{Total}_{t-1}}{\Sigma}. \tag{3.7}$$

Then one needs to update the value of Total (to be used in the computation of time $t+1$ value of LMA)

$$\text{Total}_t = \text{Total}_{t-1} + P_t - P_{t-n}. \tag{3.8}$$

Whereas the calculation of LMA according to formula (3.6) requires $2n - 1$ operations ($n-1$ multiplications, $n-1$ summations, and one division), the calculation of LMA according to formulas (3.7) and (3.8) requires 6 operations regardless of the size of the averaging window (one multiplication, one division, two summations, and two subtractions).

The average lag time of LMA is given by (see the subsequent appendix for the details of the derivation)

$$\text{Lag time}(LMA_n) = \frac{\sum_{i=1}^{n-1}(n-i) \times i}{\sum_{i=0}^{n-1}(n-i)} = \frac{n-1}{3}. \tag{3.9}$$

Notice that the average lag time of $LMA(n)$ amounts to 2/3 of the average lag time of $SMA(n)$. Consequently, for the same size of the averaging window, the lag time of LMA is smaller than that of SMA; this is illustrated in Fig. 3.1, top panel. Specifically, the plot in this panel demonstrates the values of $SMA(16)$ and $LMA(16)$ computed using the monthly closing prices of the S&P 500 index over a 10-year period from January 1997 to December 2006. This specific historical period is chosen for illustrations because over this period the trend in the S&P 500 index is clear-cut with two major turning points. Between the turning points, the index moves steadily upward or downward. Apparently, $LMA(16)$ lags behind the S&P 500 index with a shorter delay than $SMA(16)$. Observe that most of the time both $LMA(16)$ and $SMA(16)$ move parallel. This behavior is due to the result, established in Sect. 2.3, which says that when prices trend steadily, all moving averages move parallel in a graph.

However, if analysts want a moving average with a smaller lag time, instead of using LMA they can alternatively decrease the window size in SMA. Therefore, a fair comparison of the properties of the two moving averages requires using LMA and SMA with the same lag time. The bottom panel in Fig. 3.1 shows the values of $SMA(11)$ and $LMA(16)$ computed using the same stock index values. Both of the moving averages have the same lag time of 5 (months). Rather surprisingly, contrary to the common belief that these two types of moving averages are inherently different, both of them move close together.

To illustrate the source of confusion and help explain why SMA and LMA with the same average lag time are very similar, Fig. 3.2, left panel, plots the price weighting functions of $SMA(11)$ and $LMA(16)$. Obviously, the two price weighting functions are intrinsically different because seemingly each price lag contributes generally very differently to the value of a moving average. This gives rise to the belief that the values of these two moving averages differ a lot. In contrast, the right panel in the same figure plots the price-change weighting functions of $SMA(11)$ and $LMA(16)$, and both of these weighting functions look essentially similar. In particular, the differences in the two price-change weighting functions are marginal. This helps explain why the two moving averages move largely right together. Since a price-change weighting function shows the contribution of each price-change lag to the value of a moving

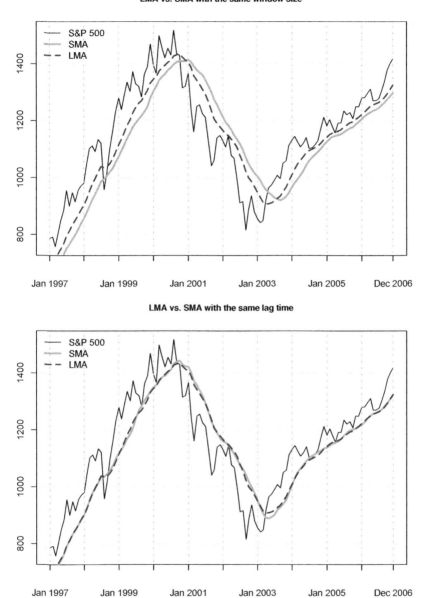

Fig. 3.1 LMA and SMA applied to the monthly closing prices of the S&P 500 index

average, it could be argued that a price-change weighting function represents the dynamic properties of a moving average. Therefore it could be argued further that a price-change weighting function provides a much more relevant information about the properties of a moving average than the corresponding price weighting function.

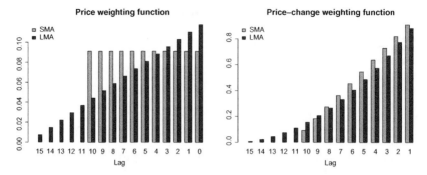

Fig. 3.2 Weighting functions of LMA and SMA with the same lag time of 5 periods. In the price weighting functions, Lag i denotes the lag of P_{t-i}. In the price-change weighting functions, Lag j denotes the lag of ΔP_{t-j}

Another reason, for why $SMA(11)$ and $LMA(16)$ largely move together in the bottom panel of Fig. 3.1, is the result established in Sect. 2.3. This result says that, when prices move steadily upward or downward, all types of moving averages with the same average lag time have basically the same values. Therefore, they move virtually together as a single moving average. The differences between different types of moving averages with the same lag time appear most often during the periods of their adaptation to the changes in the trend.

The Herfindahl index of LMA is given by (see the subsequent appendix for the details of the derivation)

$$HI(LMA_n) = \frac{2}{3} \times \frac{(2n+1)}{n(n+1)}. \tag{3.10}$$

For a sufficiently large size of the averaging window,[1]

$$HI(LMA_n) \approx \frac{4}{3} \times \frac{1}{n} = \frac{4}{3} \times HI(SMA_n). \tag{3.11}$$

That is, for the same size of the averaging window, LMA has not only smaller lag time than that of SMA, but also lower smoothness. Combining Eqs. (3.9) and (3.11) yields

$$\text{Lag time}(LMA_n) \approx \frac{4}{9} \times \text{Smoothness}(LMA_n) - \frac{1}{3}. \tag{3.12}$$

---

[1]More formally, when $n \gg 1$.

As for SMA, the average lag time of LMA is a linear function of its smoothness. The comparison of Eqs. (3.4) and (3.12) reveals that, when the value of smoothness > 3, for the same smoothness LMA has smaller average lag time than SMA. Similarly, for the same lag time LMA has higher smoothness than SMA. Therefore, for example, $SMA(11)$ and $LMA(16)$ have the same average lag time, but $LMA(16)$ is a bit smoother than $SMA(11)$.

To further highlight the difference between SMA and LMA with the same average lag time, we apply $SMA(11)$ and $LMA(16)$ to the trend component T_t of the artificial stock price data considered in Chap. 1. We remind the reader that this artificial stock price trend is given by two linear segments. First, the stock price trends upward, then downward. Our goal is to visualize the behavior of $SMA(11)$ and $LMA(16)$ along the stock price trend in general, and their reactions to a sharp change in the trend in particular. The illustration is provided in Fig. 3.3. As expected, both $SMA(11)$ and $LMA(16)$ generally move together; yet there are marginal but noticeable differences in the values of the two moving averages around their tops. Specifically, when the prices are trending, both of the moving averages lag behind the trend by 5 periods. However, while the turning point in $SMA(11)$ lags behind the turning point

SMA and LMA with the same lag time

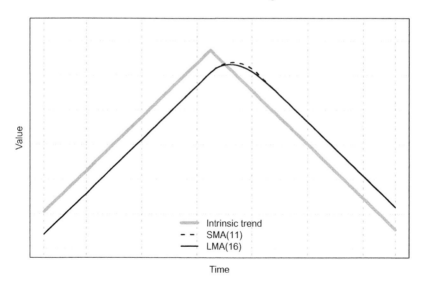

Intrinsic trend

SMA(11)

LMA(16)

Time

Fig. 3.3 Illustration of the behavior of $SMA(11)$ and $LMA(16)$ along the stock price trend and their reactions to a sharp change in the trend. Note that when prices trend upward or downward, the values of the two moving averages coincide. The differences between the values of these two moving averages appear during the period of their "adaptation" to the change in the trend

in the trend by 5 periods, the turning point in $LMA(16)$ lags behind the turning point in the trend by 4 periods.[2] Consequently, moving averages that overweight the most recent prices may indeed possess advantages over the equally-weighted moving average. These advantages consist not only in better smoothness for the same average lag time, but also in earlier detection of turning points in a trend. Therefore in market timing applications LMA might have a potential advantage over SMA. Yet even in ideal conditions, without an additive noise component, this advantage is marginal. The presence of noise can totally nullify this advantage.

3.1.3 Exponential Moving Average

A disadvantage of the linearly weighted moving average is that its weighting scheme is too rigid. This problem can be addressed by using the exponentially weighted moving average instead of the linearly weighted moving average. An Exponential Moving Average (EMA) is computed as

$$EMA_t(\lambda, n) = \frac{P_t + \lambda P_{t-1} + \lambda^2 P_{t-2} + \cdots + \lambda^{n-1} P_{t-n+1}}{1 + \lambda + \lambda^2 + \cdots + \lambda^{n-1}} = \frac{\sum_{i=0}^{n-1} \lambda^i P_{t-i}}{\sum_{i=0}^{n-1} \lambda^i},$$
(3.13)

where $0 < \lambda \leq 1$ is a decay factor. When $\lambda < 1$, the exponentially weighted moving average assigns greater weights to the most recent prices. By varying the value of λ, one is able to adjust the weighting to give greater or lesser weight to the most recent price. The properties of the exponential moving average are as follows:

$$\lim_{\lambda \to 1} EMA_t(\lambda, n) = SMA_t(n), \quad \lim_{\lambda \to 0} EMA_t(\lambda, n) = P_t. \quad (3.14)$$

In words, when λ approaches unity, the value of EMA converges to the value of the corresponding SMA. When λ approaches zero, the value of EMA becomes the last closing price.

The average lag time of EMA is given by (see the subsequent appendix for the details of the derivation)

$$\text{Lag time}(EMA_{\lambda, n}) = \frac{\lambda - \lambda^n}{(1 - \lambda)(1 - \lambda^n)} - \frac{(n-1)\lambda^n}{1 - \lambda^n}. \quad (3.15)$$

---

[2]The delay in the identification of the turning point in a trend is estimated numerically as the time difference between the maximum value of the price and the maximum value of the moving average.

The average lag time of EMA depends on the value of two parameters: the decay factor λ and the size of the averaging window n. For example, to reduce the average lag time, one can either reduce the window size n or decrease the decay factor λ. Consequently, there are infinitely many combinations of $\{\lambda, n\}$ that produce EMAs with exactly the same average lag time; at the same time these moving averages have similar type of the weighting function. As a result, these EMAs possess basically similar properties.

To get rid of the unwarranted redundancy in the parameters of the EMA with a finite size of the averaging window, analysts use EMA with an infinite size of the averaging window. Specifically, analysts compute EMA as

$$EMA_t(\lambda) = \frac{P_t + \lambda P_{t-1} + \lambda^2 P_{t-2} + \lambda^3 P_{t-3} + \cdots}{1 + \lambda + \lambda^2 + \lambda^3 + \cdots} = (1-\lambda) \sum_{i=0}^{\infty} \lambda^i P_{t-i},$$

$$(3.16)$$

where the last equality follows from the fact that $\sum_{i=0}^{\infty} \lambda^i = (1-\lambda)^{-1}$. For an infinite EMA, the average lag time is given by

$$\text{Lag time}(EMA_\lambda) = \frac{\lambda}{1-\lambda}, \qquad (3.17)$$

which is obtained as a limiting case of the average lag time of a finite EMA (given by Eq. (3.15)) when $n \to \infty$.

Even though an infinite EMA is free from the redundancy of a finite EMA, using EMA together with the other types of moving averages is inconvenient because the key parameter of EMA is the decay constant, whereas in both SMA and LMA the key parameter is the size of the averaging window. To unify the usage of all types of moving averages, analysts also use the size of the averaging window as the key parameter in the (infinite) EMA. The idea is that EMA with the window size of n should have the same average lag time as SMA with the same window size. Equating the average lag time of $SMA(n)$ with the average lag time of $EMA(\lambda)$ gives

$$\frac{n-1}{2} = \frac{\lambda}{1-\lambda}.$$

The solution of this equation with respect to λ yields

$$\lambda = \frac{n-1}{n+1}. \qquad (3.18)$$

As a result, EMA is computed according to the following formula:

$$EMA_t(n) = (1 - \lambda) \sum_{i=0}^{\infty} \lambda^i P_{t-i}, \text{ where } \lambda = \frac{n-1}{n+1}. \tag{3.19}$$

The formula for EMA can be rewritten in the following manner

$$EMA_t(n) = (1 - \lambda) P_t + \lambda(1 - \lambda) \sum_{i=0}^{\infty} \lambda^i P_{t-1-i}.$$

Since

$$(1 - \lambda) \sum_{i=0}^{\infty} \lambda^i P_{t-1-i} = EMA_{t-1}(n),$$

the formula for EMA can be written in a recursive form that can greatly facilitate and accelerate the computation of EMA in practice

$$EMA_t(n) = (1 - \lambda) P_t + \lambda \, EMA_{t-1}(n). \tag{3.20}$$

In the formula above, $(1 - \lambda)$ determines the weight of the last closing price in the computation of the current EMA, whereas λ determines the weight of the previous EMA in the computation of the current EMA.

In practice, it is more common to write the recursive formula for EMA using parameter

$$\alpha = 1 - \lambda.$$

The recursive formula for EMA is usually written therefore as

$$EMA_t(n) = \alpha \, P_t + (1 - \alpha) EMA_{t-1}(n),$$

and the value of the parameter α, in terms of the window size of SMA with the same average lag time, is given by

$$\alpha = \frac{2}{n+1}. \tag{3.21}$$

Notice that the larger the window size n, the smaller the parameter α. That is, when n increases, the weight of the last closing price in the current EMA decreases while the weight of the previous EMA increases. For example, if $n = 9$, the value of α equals 0.2 or 20%. Consequently, in the 9-day EMA

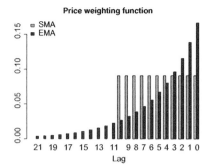
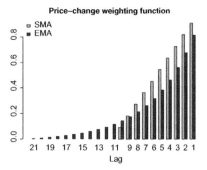

Fig. 3.4 Weighting functions of EMA and SMA with the same lag time of 5 periods. In the price weighting functions, Lag i denotes the lag of P_{t-i}. In the price-change weighting functions, Lag j denotes the lag of ΔP_{t-j}. The weights of the (infinite) EMA are cut off at lag 21

the weight of the last closing price amounts to 20%, while the weight of the previous EMA equals 80%. If, on the other hand, $n = 19$, the value of α equals 10%. Thus, in the 19-day EMA the weight of the last closing price amounts to 10%, while the weight of the previous EMA equals 90%.

Figure 3.4 plots the price weighting functions (left panel) and the price-change weighting functions (right panel) of $SMA(11)$ and $EMA(11)$. For EMA, not only the price weighting function is substantially different from that of SMA, but there are also notable (yet not very significant) differences between the two price-change weighting functions.

Figure 3.5 plots the values of $SMA(11)$ and $EMA(11)$ computed using the monthly closing prices of the S&P 500 index over a 10-year period from January 1997 to December 2006. Both of the moving averages have the same lag time of 5 (months). The plot in this figure suggests that the values of $SMA(11)$ and $EMA(11)$ move close together when the stock prices trend upward or downward. This comes as no surprise given the previously established fact that all moving averages with different weighting functions but the same average lag time move close together when the trend is strong. Only when the direction of trend is changing, we see that the values of $SMA(11)$ and $EMA(11)$ start to move slightly apart. The plot in this figure motivates that, when the direction of trend is changing, EMA follows the trend more closely than SMA. Therefore, EMA might have a potential advantage over SMA with the same average lag time.

The Herfindahl index of the infinite EMA is given by (see the subsequent appendix for the details of the derivation)

$$HI(EMA_n) = \frac{1}{n} = HI(SMA_n). \tag{3.22}$$

EMA vs. SMA with the same lag time

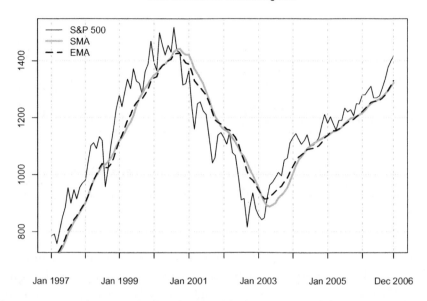

Fig. 3.5 EMA and SMA applied to the monthly closing prices of the S&P 500 index

That is, not only the average lag time of $EMA(n)$ equals the average lag time of $SMA(n)$, but also the smoothness of both these moving averages is alike (at least in theory).

To further highlight the difference between SMA and EMA with equal average lag times, we apply $SMA(11)$ and $EMA(11)$ to the same artificial stock price trend as in the preceding section. The illustration is provided in Fig. 3.6. Both of these moving averages have the same average lag time of 5 periods. As expected, when the prices are trending, both of the moving averages lag behind the trend by 5 periods. However, while the turning point in $SMA(11)$ lags behind the turning point in the trend by 5 periods, the turning point in $EMA(11)$ lags behind the turning point in the trend by 3 periods. Consequently, EMA might have a potential advantage over both SMA and LMA with the same average lag time.[3]

---

[3]Yet recall that $LMA(16)$, which has the same average lag time as that of $EMA(11)$, has slightly better smoothing properties than those of $EMA(11)$. Also keep in mind that the delay in turning point detection is evaluated using a specific artificial stock price trend with one turning point. Therefore the result on the lag time in turning point identification cannot be generalized for all types of trend changes.

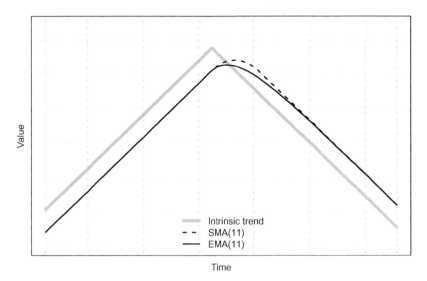

SMA and EMA with the same lag time

Fig. 3.6 Illustration of the behavior of $SMA(11)$ and $EMA(11)$ along the stock price trend and their reactions to a sharp change in the trend

3.2 Moving Averages of Moving Averages

Increasing the size of the averaging window is not the only way to improve smoothing properties of a moving average. Another possibility is to smooth a moving average by another moving average. The result of this operation is a new moving average which is usually called a "double moving average". A double moving average can itself be smoothed further by another moving average producing a "triple moving average". Such an iterative smoothing can be repeated a number of times, if desired.

In the rest of this chapter, in order to simplify the notation, we will denote by $MA_n(X)$ (or just by $MA(X)$) a moving average of a time series X computed using the window size of n. Using this notation, $MA_n(MA_n(P))$ denotes a double moving average, whereas $MA_n(MA_n(MA_n(P)))$ denotes a triple moving average of a series of prices.

3.2.1 Triangular Moving Average

A Triangular Moving Average (TMA) is a simple moving average of prices smoothed by another simple moving average with the same size of the averaging window:

$$TMA_t(m) = SMA_n(SMA_n(P)).$$

Specifically,

$$TMA_t(m) = \frac{SMA_t(n) + SMA_{t-1}(n) + \cdots + SMA_{t-n+1}(n)}{n} = \frac{1}{n}\sum_{i=0}^{n-1} SMA_{t-i}(n).$$

Notice that, for any moving average with a finite size n of the averaging window, a double moving average of prices is a new type of a moving average of prices computed using the window size of $m = 2n - 1$. This is because, for example, to compute $TMA_t(m)$ one needs to know the value of $SMA_{t-n+1}(n)$ which is computed using the prices $(P_{t-n+1}, \ldots, P_{t-2(n-1)})$. Thus, $TMA_t(m)$ is computed using the prices $(P_t, \ldots, P_{t-2(n-1)})$.

The average lag time of TMA is given by

$$\text{Lag time}(TMA_m) = \frac{m-1}{2} = n - 1,$$

which is twice the average lag time of a single SMA_n used to create TMA_m. Therefore, for instance, TMA_{11} and SMA_{11} have exactly the same lag time, but TMA_{11} is constructed as $SMA_6(SMA_6)$. In addition, since for a fixed size n of the averaging window the equal weighting of prices in a moving average (as in $SMA(n)$) provides the best smoothness, $TMA(n)$, that has the same window size as $SMA(n)$, has lower smoothness than that of $SMA(n)$.

Figure 3.7 plots the price weighting functions of $SMA_{11}(P)$ and $SMA_{11}(SMA_{11}(P))$. The price weighting function of the triangular moving average represents an isosceles triangle, hence the name. Figure 3.8 plots

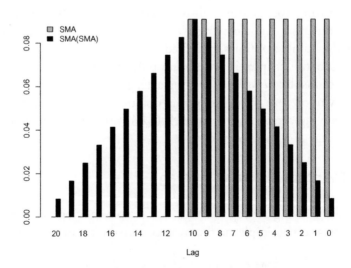

Fig. 3.7 Price weighting functions of $SMA_{11}(P)$ and $SMA_{11}(SMA_{11}(P))$

TMA vs. SMA with the same lag time

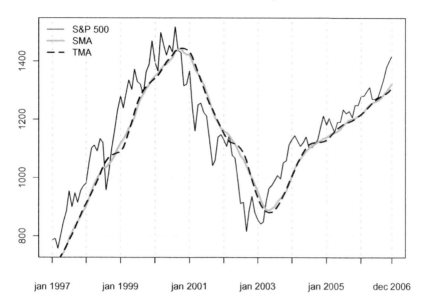

Fig. 3.8 $TMA_{11}(P)$ and $SMA_{11}(P)$ applied to the monthly closing prices of the S&P 500 index

the values of $TMA_{11}(P)$ and $SMA_{11}(P)$ computed using the monthly clos-ing prices of the S&P 500 index over a 10-year period from January 1997 to December 2006. Both of the moving averages have the same average lag time of 5 (months). The visual inspection of the two moving averages suggests that the differences between them are marginal; they move really close together. In addition, both the moving averages have about the same delay in the detection of turning points.

3.2.2 Double and Triple Exponential Smoothing

Double and triple exponential smoothing is the recursive application of EMA two and three times respectively. Figure 3.9 plots the price weight-ing functions of $EMA_{11}(P)$, $EMA_{11}(EMA_{11}(P))$, and $EMA_{11}(EMA_{11}(EMA_{11}(P)))$. As the reader may note, the recursive application of a mov-ing average changes the price weighting function of a moving average. In case of a moving average with a finite size of the averaging window, the recursive smoothing decreases the weights of the most recent and the most distant data (as in TMA). In case of an infinite EMA which heavily overweights the most recent data, the recursive smoothing decreases the weights of the most recent data. As a consequence, after a recursive smoothing the price weighting func-

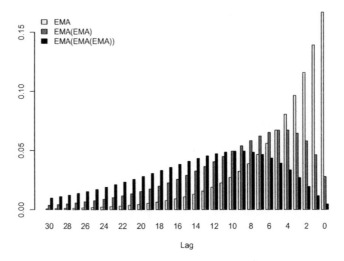

Fig. 3.9 Price weighting functions of $EMA_{11}(P)$, $EMA_{11}(EMA_{11}(P))$, and $EMA_{11}(EMA_{11}(EMA_{11}(P)))$. The weights of the (infinite) EMAs are cut off at lag 30

tion of the resulting moving average acquires a hump-shaped form. This price weighting function underweights both the most recent and most distant data.

In addition, the recursive smoothing increases the average lag time of a moving average. Specifically,

$$\text{Lag time}(EMA_n(EMA_n)) = n - 1,$$

which is double the lag time of EMA_n. Further,

$$\text{Lag time}(EMA_n(EMA_n(EMA_n))) = \frac{3}{2}(n - 1),$$

which is triple the lag time of EMA_n. Last but not least, since the recursive smoothing decreases the weights of the most recent prices (as compared with the weighting function of EMA), the recursive smoothing increases the delay in the detection of turning points (again, as compared with that of EMA).

3.3   Mixed Moving Averages with Less Lag Time

The smoothness of a moving average is generally inversely related to its average lag time. That is, as a rule, the better the smoothness of a moving average is, the large its average lag time. There have been many attempts to improve the tradeoff between the smoothness and the average lag time of a moving average. Some of the examples of moving averages with less average lag time are considered in this section. The common feature of these moving averages is that the price weighting functions of these moving averages assign negative weights to more distant prices in the averaging window.

Specifically, consider the computation of the average lag time of a moving average given by Eq. (2.4). The average lag time is computed as the weighted average "age" of data used to compute the moving average of prices. If one allows negative weights in the price weighting function of a moving average, one can reduce the average lag time to zero. In principle, one can make the average lag time to be even negative. In this case it may seem that a moving average, instead of being a lagging indicator, becomes miraculously a leading indicator and can easily predict the direction of the future stock price trend. Unfortunately, miracles do not happen in the real world. In this context, it is worth repeating that only in cases where the prices are steadily increasing (or decreasing) over a relatively long period of time, the "average lag time" provides a correct numerical characterization of the time lag between the price and the value of the moving average.

In practical applications, a much more relevant characteristic of the properties of a moving average is its lag time in the detection of turning points in a price trend. Using negative weights in the price weighting function of a moving average does not allow one to predict turning points in a trend; turnings points in a trend can be identified only a posteriori; this will be illustrated shortly. In addition, by means of an example, we will also show that moving averages with negative weights in the price weighting function, that have the same delay in turning point identification as that of the respective ordinary moving averages, have worse smoothing properties. Therefore these moving averages tend to deteriorate the tradeoff between the smoothness and the lag time in turning point identification.

3.3.1 Zero Lag Exponential Moving Average

This type of a moving average was suggested by Ehlers and Way (2010). The idea behind the construction of their Zero Lag Exponential Moving Average (ZLEMA) is as follows. The regular $EMA_t(n, P)$ has the average lag time of $\frac{n-1}{2}$ and its value differs from the value of the last closing price P_t due to the lagging nature of the moving average. The discrepancy between the last closing price and the value of $EMA_t(n, P)$ can be estimated as (for motivation, see Fig. 2.1)

$$P_t - Lag_{\frac{n-1}{2}}(P_t),$$

where Lag_j is the lag operator defined by

$$Lag_j(P_t) = P_{t-j}.$$

In words, $Lag_j(P_t)$ is the value of the time series of prices at time $t - j$.

To push the value of the moving average closer towards the value of the last closing price, one possibility is to add the estimated discrepancy to the value of the moving average

$$EMA_n(P) + \left(P - Lag_{\frac{n-1}{2}}(P)\right).$$

However, because the price is noisy, in this case the resulting combination loses smoothness. The solution proposed by Ehlers and Way (2010) is to smooth both the price and the estimated discrepancy:

$$ZLEMA = EMA_n\left(P + \left(P - Lag_{\frac{n-1}{2}}(P)\right)\right). \qquad (3.23)$$

Since any moving average is a linear operator (see Eq. (2.3)), the formula for ZLEMA can be rewritten as

$$ZLEMA = EMA_n\left(2P - Lag_{\frac{n-1}{2}}(P)\right) = 2 \times EMA_n(P) - EMA_n\left(Lag_{\frac{n-1}{2}}(P)\right).$$

Therefore ZLEMA can be considered as a (linear) combination of two EMAs.

Figure 3.10 plots the price weighting function of ZLEMA as well as the price weighting function of EMA_{11} used to create this ZLEMA. Notice that in ZLEMA the weights of the price lags from 5 and beyond are negative. Figure 3.11 plots the values of $ZLEMA(P)$ and $EMA_{11}(P)$ computed using the monthly closing prices of the S&P 500 index over a 10-year period from January 1997 to December 2006. Observe that indeed, due to the presence of negative weights in its price weighting function, ZLEMA follows the prices much more closely than the EMA used to create this ZLEMA. However, it is important to observe that at the same time ZLEMA is less smooth than EMA. Specifically, period to period variations in ZLEMA are greater than those in EMA. Whereas the Herfindahl index of EMA_{11} equals $\frac{1}{11} = 0.091$, the Herfindahl index of ZLEMA, based on EMA_{11}, equals 0.308 (the latter index is computed numerically using ZLEMA weights).

Despite the fact that ZLEMA has almost zero average lag time, ZLEMA identifies turning points in a trend with delay. To illustrate this, we apply

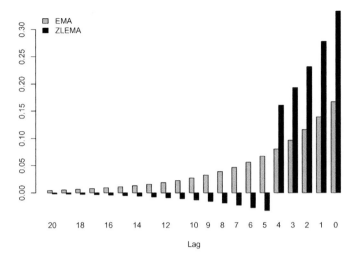

Fig. 3.10 Price weighting functions of EMA_{11} and ZLEMA based on EMA_{11}. The weights of the (infinite) EMAs are cut off at lag 21

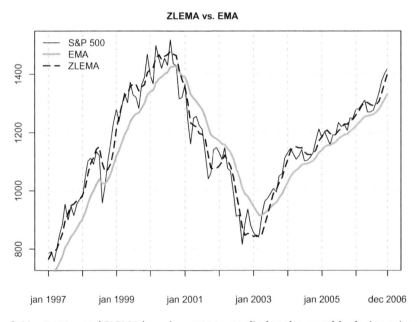

Fig. 3.11 EMA_{11} and ZLEMA based on EMA_{11} applied to the monthly closing prices of the S&P 500 index

EMA_{11} and ZLEMA to the same artificial stock price trend as in the preceding sections. Our goal in this exercise is to find the lag time of EMA used to construct ZLEMA such that the resulting ZLEMA and EMA_{11} have the

ZLEMA and EMA with the same delay in turning point identification

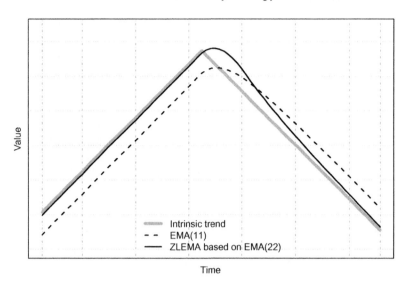

Fig. 3.12 Illustration of the behavior of EMA_{11} and ZLEMA (based on EMA_{22}) along the stock price trend and their reactions to a sharp change in the trend. Both EMA_{11} and ZLEMA have the same lag time of 3 periods in the detection of the turning point in the trend

same lag time in the identification of the turning point in the artificial stock price trend. Previously, we estimated that EMA_{11} identifies the turning point in the artificial trend with a delay of 3 periods. We find that ZLEMA based on EMA_{22} has the same 3-period delay in the turning point detection. The illustration of this result is provided in Fig. 3.12. Notice that, when the prices are trending, ZLEMA (based on EMA_{22}) follows the trend with almost zero lag, whereas EMA_{11} has the lag time of 5 periods. However, when the direction of the trend is sharply changing, ZLEMA needs some time to adapt to the new direction of the trend. During this "adaptation period", ZLEMA lags behind the trend. Consequently, this illustration demonstrates that ZLEMA has almost zero lag time only when prices are trending steadily over a relatively long period. Last but not least, ZLEMA based on EMA_{22} still has a higher Herfindahl index of 0.154. That is, both EMA_{11} and ZLEMA based on EMA_{22} have the same delay in the identification of the turning point, yet EMA_{11} is smoother than ZLEMA. Therefore it is doubtful that ZLEMA possesses any potential advantages over EMA in practical applications.

3.3.2   Double and Triple Exponential Moving Average

A Double Exponential Moving Average (DEMA) is a mixed moving average proposed by Mulloy (1994a). The original idea of Mulloy was to reduce the lag time of the regular EMA by placing more weight (than in regular EMA) on the most recent prices. The value of DEMA is computed according to the following formula

$$DEMA = 2 \times EMA_n(P) - EMA_n(EMA_n(P)). \tag{3.24}$$

To understand why DEMA has very small average lag time, using the linearity property of moving averages we rewrite the formula for DEMA as

$$DEMA = EMA_n(2P - EMA_n(P)) = EMA_n\big(P + (P - EMA_n(P))\big).$$

In this form, it becomes apparent that DEMA exploits the same idea as that in ZLEMA. In particular, in order to reduce the average lag time, DEMA pushes the value of the moving average closer towards the value of the last closing price. While ZLEMA uses for this purpose the estimated discrepancy between P_t and $EMA_t(n, P)$, DEMA uses the exact discrepancy between P_t and $EMA_t(n, P)$.

Subsequently, Mulloy (1994b) proposed a Triple Exponential Moving Average (TEMA) with even less average lag time as that of DEMA. The value of TEMA is computed according to

$$TEMA = 3 \times EMA_n(P) - 3 \times EMA_n(EMA_n(P)) + EMA_n(EMA_n(EMA_n(P))). \tag{3.25}$$

Using the linearity property of moving averages, we can rewrite the formula for TEMA as

$$TEMA = EMA_n\Big(P + 2\big(P - EMA_n(P)\big) - EMA_n\big(P - EMA_n(P)\big)\Big).$$

That is, to reduce the average lag time, TEMA adds the double discrepancy (between P_t and $EMA_t(n, P)$) to the last closing price, subtracts the smoothed value of this discrepancy, and performs exponential smoothing of the resulting time series.

Figure 3.13, left panel, plots the price weighting function of DEMA as well as the price weighting function of EMA_{11} used to create this DEMA. Similarly, Fig. 3.13, right panel, plots the price weighting function of TEMA as well as the price weighting function of EMA_{11} used to create this TEMA. Figure 3.14 plots the values of $EMA_{11}(P)$, $DEMA(P)$, and

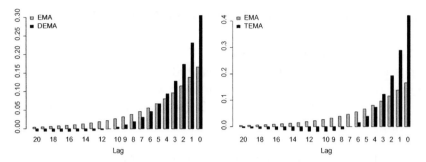

Fig. 3.13 Price weighting functions of EMA_{11}, DEMA based on EMA_{11}, and TEMA based on EMA_{11}. The weights of the (infinite) EMAs are cut off at lag 21

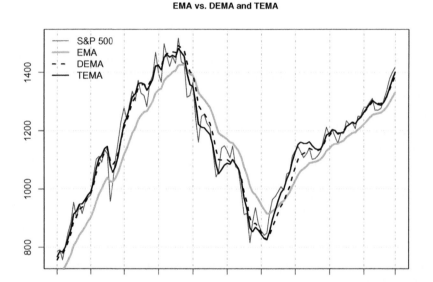

Fig. 3.14 EMA_{11}, DEMA and TEMA (both of them are based on EMA_{11}) applied to the monthly closing prices of the S&P 500 index

$TEMA(P)$ computed using the monthly closing prices of the S&P 500 index over a 10-year period from January 1997 to December 2006.

Using a numerical method, we estimate that EMA_{11}, DEMA based on EMA_{22}, and TEMA based on EMA_{30} have the same delay in the identification of the turning point in the artificial stock price trend. However, whereas the Herfindahl index of EMA_{11} equals 0.091, the Herfindahl index of DEMA based on EMA_{22} equals 0.110 and the Herfindahl index of TEMA based on EMA_{30} equals 0.130. That is, both DEMA and TEMA have worse smoothness than EMA with a comparable delay in the identification of turning points.

3.3.3 Hull Moving Average

To reduce the average lag time, Hull (2005) proposed a combination of 3 LMAs with different sizes of the averaging window. The Hull Moving Average (HMA) is computed as

$$HMA = LMA_{\sqrt{n}}\left(2 \times LMA_{\frac{n}{2}}(P) - LMA_n(P)\right). \qquad (3.26)$$

HMA is constructed using basically the same idea as that used for the construction of ZLEMA and DEMA. Specifically, a general method for the construction of a moving average with less average lag time can be described by the following formula

$$MA_n\left(2 \times MA_s(P) - MA_l(P)\right) = MA_n\left(MA_s(P) + \left(MA_s(P) - MA_l(P)\right)\right)$$

where MA denotes a moving average, s denotes the size of a short averaging window, l denotes the size of a long averaging window (such that $l > s$), and n denotes the size of the averaging window for final smoothing. That is, to construct a moving average with less average lag time, one performs a gentle (or no) smoothing of the price series using a short window s, adds to the result a proxy for the discrepancy between the result and the last closing price, and finally smoothes the aggregate time series. Observe that when $MA = EMA$, $s = 1$, and $l = n$, then this general method describes the computation of DEMA. If, in addition, one uses the lagged price series $Lag_l(P)$ instead of $MA_l(P)$, then this method describes the computation of ZLEMA.

Figure 3.15 plots the price weighting function of LMA_{16} as well as the price weighting function of HMA based on using the same size of the averaging window of $n = 16$. Figure 3.16 plots the values of LMA_{16} and HMA (based on $n = 16$) computed using the monthly closing prices of the S&P 500 index over a 10-year period from January 1997 to December 2006.

Using a numerical method, we estimate that LMA_{16} and HMA based on $n = 28$ have the same delay in the identification of the turning point in the artificial stock price trend. However, whereas the Herfindahl index of LMA_{16} equals 0.081, the Herfindahl index of HMA based on $n = 28$ equals 0.152. That is, HMA has worse smoothness than LMA with a comparable delay in the identification of turning points.

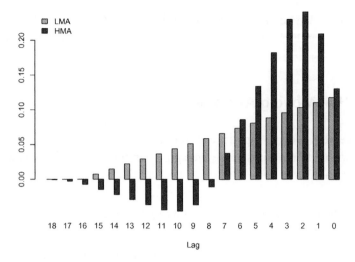

Fig. 3.15   Weighting functions of LMA_{16} and HMA based on $n = 16$

HMA vs. LMA

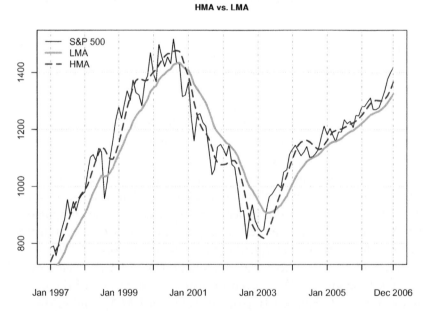

Fig. 3.16 LMA_{16} and HMA (based on $n = 16$) applied to the monthly closing prices of the S&P 500 index

3.4   Chapter Summary

The two important characteristics of a moving average are the lag time and smoothness. Analysts want a moving average to have short lag time and high smoothness. This is because the shorter the lag time is, the earlier turning points in a trend can be recognized. The trading frequency in a market timing strategy is inversely related to the smoothness of a moving average. Using a less smooth moving average results in a larger number of trades and, consequently, in larger transaction costs. In addition, using a moving average with insufficient smoothness results in generation of many false signals. Unfortunately, for each specific type of a moving average, its lag time and smoothness are directly related. That is, the less the lag time is, the worse the smoothness.

In the preceding chapter we established that each moving average is uniquely characterized by its price weighting function. The weights in this function are used to compute the average lag time and smoothness of a moving average. In this chapter we considered all ordinary moving averages used by analysts and a few exotic moving averages. The exotic moving averages include moving averages of moving averages and mixed moving averages with less lag time. Each of these moving averages (both ordinary and exotic) has a unique weighting function and, therefore, each of these moving averages provides different tradeoff between the lag time and smoothness.

We assert that the notion of the "lag time" of a moving average is an elusive concept. In the preceding chapter we argued that the quantity, known as the "average lag time" of a moving average, provides a correct numerical characterization of the time lag between the price and the value of a moving average of prices only when prices are steadily increasing or decreasing. Our analysis reveals that there are two issues with the notion of the "average lag time". First, the average lag time has little to do with the delay in the identification of turning points in a trend. Second, the average lag time can be easily reduced (that is, manipulated) by using a weighting function with negative weights.

Using an artificial stock price trend with one turning point, we demonstrated that moving averages that overweight the most recent prices provide a better tradeoff between the smoothness and the delay in turning point identification than that provided by the moving average with equal weighting of prices. That is, our illustration suggests that LMA and EMA have some potential advantages over SMA. Using the same artificial stock price trend, we also demonstrated that moving averages with reduced (by means of using negative weights in a weighting function) average lag time have worse tradeoff between the smoothness and the delay in turning point identification than that provided by the ordinary moving averages. Unfortunately, these conclusions cannot be generalized because they were drawn based on a particular example. In each specific case the delay in turning point identification depends not only on the price weighting function of a moving average, but also on the strengths of the trend before and after the turning point and on the amount of noise in the price series.

Appendix 3.A: Formulas for Sums of Sequences and Series

3.A.1 Sequence

A sequence is a set of numbers that are in order. Denote the n-th term of a sequence by a_n. Then the sequence is given by

$$\{a_1, a_2, a_3, \ldots, a_n, \ldots\}.$$

3.A.2 Arithmetic Sequence

An arithmetic sequence is a sequence of numbers where each term is found by adding a constant (called the "common difference") to the previous term. If the initial term of an arithmetic sequence is a_1 and the common difference is d, then the n-th term of the sequence is given by

$$a_n = a_1 + (n - 1) \times d.$$

The sum of the first n terms of an arithmetic sequence is given by

$$S_n = \sum_{i=1}^{n} a_i = \frac{n(a_1 + a_n)}{2}. \tag{3.27}$$

3.A.3 Geometric Sequence

In a geometric sequence each term is found by multiplying the previous term by a constant (called the "common ratio"). If the initial term of a geometric sequence is a and the common ratio is r, then the n-th term of the sequence is given by

$$a_n = a \times r^{n-1}.$$

The sum of the first n terms of a geometric sequence is given by

$$S_n = \sum_{i=1}^{n} a_i = a \left(\frac{1 - r^n}{1 - r} \right). \tag{3.28}$$

If $0 < r < 1$, then the sum of the infinite geometric sequence

$$S_\infty = \sum_{i=1}^{\infty} a_i = \frac{a}{1-r}. \tag{3.29}$$

3.A.4 Sequence of Squares

A sequence of squares is given by

$$\{1^2, 2^2, 3^2, \ldots, n^2, \ldots\}.$$

The sum of the first n terms of a sequence of squares is given by

$$S_n = \sum_{i=1}^{n} i^2 = \frac{n(n+1)(2n+1)}{6}. \tag{3.30}$$

Appendix 3.B: Derivation of Formulas for Lag Times and Herfindahl indices of Some Moving Averages

3.B.1 Average Lag Time of SMA

Start with

$$\text{Lag time}(SMA_n) = \frac{\sum_{i=1}^{n-1} i}{\sum_{i=0}^{n-1} 1} = \frac{1 + 2 + \cdots + (n-1)}{n}.$$

The numerator in the fraction above is the sum of the first $n - 1$ terms of an arithmetic series with $a_1 = 1$ and $d = 1$. This sum is given by $\frac{n(n-1)}{2}$. Therefore

$$\text{Lag time}(SMA_n) = \frac{\frac{n(n-1)}{2}}{n} = \frac{n-1}{2}.$$

3.B.2 Average Lag Time and Herfindahl Index of LMA

The average lag time of LMA is computed according to

$$\text{Lag time}(LMA_n) = \frac{\sum_{i=1}^{n-1}(n-i) \times i}{\sum_{i=0}^{n-1}(n-i)} = \frac{n\sum_{i=1}^{n-1}i - \sum_{i=1}^{n-1}i^2}{\sum_{i=1}^{n}i}.$$

We need to derive closed-form expressions for three sums in this formula, where two of them are sums of arithmetic sequences and one of them is a sum of a sequence of squares. The derivations give

$$\sum_{i=1}^{n}i = \frac{n(n+1)}{2}, \quad \sum_{i=1}^{n-1}i = \frac{n(n-1)}{2}, \quad \sum_{i=1}^{n-1}i^2 = \frac{n(n-1)(2n-1)}{6}.$$

Putting all this together gives

$$\text{Lag time}(LMA_n) = \frac{\frac{n(n-1)n}{2} - \frac{n(n-1)(2n-1)}{6}}{\frac{n(n+1)}{2}} = \frac{n-1}{3}.$$

The Herfindahl index of LMA is computed according to

$$HI(LMA_n) = \frac{\sum_{i=0}^{n-1}(n-i)^2}{\left(\sum_{i=0}^{n-1}(n-i)\right)^2} = \frac{\sum_{i=1}^{n}i^2}{\left(\sum_{i=1}^{n}i\right)^2}.$$

The formula for the sum in the denominator of this fraction is derived above. The formula for the sum in the numerator is given by (3.30). Therefore

$$HI(LMA_n) = \frac{\frac{n(n-1)(2n-1)}{6}}{\left(\frac{n(n+1)}{2}\right)^2} = \frac{2}{3} \times \frac{(2n+1)}{n(n+1)}.$$

3.B.3 Average Lag Time and Herfindahl Index of EMA

The average lag time of EMA is computed according to

$$\text{Lag time}(EMA_n) = \frac{\sum_{i=1}^{n-1}\lambda^i \times i}{\sum_{i=0}^{n-1}\lambda^i}.$$

The denominator in the fraction above is the sum of the first n terms of a geometric series with $a = 1$ and $r = \lambda$. This sum is given by $\frac{1-\lambda^n}{1-\lambda}$. Remains to derive the closed-form expression for the sum in the numerator:

$$\sum_{i=1}^{n-1} \lambda^i \times i = \lambda + 2\lambda^2 + 3\lambda^3 + \cdots + (n-1)\lambda^{n-1} = \sum_{i=1}^{n-1}\sum_{j=i}^{n-1} \lambda^j = \sum_{i=1}^{n-1} \frac{\lambda^i - \lambda^n}{1-\lambda}$$

$$= \frac{1}{1-\lambda}\left(\sum_{i=1}^{n-1}\lambda^i - \sum_{i=1}^{n-1}\lambda^n\right) = \frac{1}{1-\lambda}\left(\frac{\lambda-\lambda^n}{1-\lambda} - (n-1)\lambda^n\right).$$

The final expression for the average lag time

$$\text{Lag time}(EMA_n) = \frac{\frac{1}{1-\lambda}\left(\frac{\lambda-\lambda^n}{1-\lambda} - (n-1)\lambda^n\right)}{\frac{1-\lambda^n}{1-\lambda}} = \frac{\lambda-\lambda^n}{(1-\lambda)(1-\lambda^n)} - \frac{(n-1)\lambda^n}{1-\lambda^n}.$$

The weighting function of the infinite EMA is given by

$$\psi_i = \alpha\lambda^i, \quad \alpha = 1 - \lambda, \quad i \in \{0, 1, 2, \ldots\}.$$

The Herfindahl index of the infinite EMA is computed according to

$$HI(EMA_\infty) = \sum_{i=0}^{\infty} \psi_i^2 = \sum_{i=0}^{\infty} \alpha^2\lambda^{2i}.$$

The sum of this infinite geometric sequence is computed according to Eq. (3.29) with $a = \alpha^2$ and $r = \lambda^2$. Therefore

$$HI(EMA_\infty) = \frac{\alpha^2}{1-\lambda^2} = \frac{\alpha}{1+\lambda}.$$

Since in practice the values of the parameters are given by

$$\alpha = \frac{2}{n+1}, \quad \lambda = \frac{n-1}{n+1},$$

the formula for the Herfindahl index of the infinite EMA becomes

$$HI(EMA_\infty) = \frac{\frac{2}{n+1}}{1+\frac{n-1}{n+1}} = \frac{2}{2n} = \frac{1}{n}.$$

References

Ehlers, J. F., & Way, R. (2010). Zero Lag (Well, Almost). *Technical Analysis of Stocks and Commodities, 28*(12), 30–35.

Hull, A. (2005). *How to reduce lag in a moving average.* http://www.alanhull.com/hull-moving-average, [Online; accessed 7-October-2016]

Mulloy, P. G. (1994a). Smoothing data with faster moving averages. *Technical Analysis of Stocks and Commodities, 12*(1), 11–19.

Mulloy, P. G. (1994b). Smoothing data with less lag. *Technical Analysis of Stocks and Commodities, 12*(2), 72–80.

Part II

Trading Rules and Their Anatomy

4

Technical Trading Rules

4.1 Trading Signal Generation

A trend following strategy is typically based on switching between the market and the cash depending on whether the market prices trend upward or downward. Specifically, when the strategy identifies that prices trend upward (downward), it generates a Buy (Sell) trading signal. A Buy signal is a signal to invest in the stocks (or stay invested in the stocks), whereas a Sell signal is a signal to sell the stocks and invest in cash (or stay invested in cash).[1] Often, a "trading rule" represents a verbal description of the trading signal generation process in a specific strategy. The technical trading rules, considered in this book, use moving averages to give specific signals. An example of a trading rule of this type is as follows: buy when the last closing price is above the 200-day simple moving average; otherwise, sell. However, there are various alternative technical trading rules based on moving averages of prices.

Formally, in each technical trading rule the generation of a trading signal is a two-step process. At the first step, the value of a technical trading indicator is computed using the past prices including the last closing price

$$\text{Indicator}_t^{TR} = f(P_t, P_{t-1}, P_{t-2}, \ldots),$$

where TR denotes the trading rule and $f(\cdot)$ denotes the function that specifies how the value of the technical trading indicator is computed. At the second step, the value of the technical indicator is translated into a trading signal.

---

[1]The other, less typical strategy, is to short the stocks when a Sell signal is generated.

© The Author(s) 2017
V. Zakamulin, *Market Timing with Moving Averages*, New Developments
in Quantitative Trading and Investment, DOI 10.1007/978-3-319-60970-6_4

In all market timing rules considered in this book, a Buy signal is generated when the value of the technical trading indicator is positive. Otherwise, a Sell signal is generated. Thus,

$$\text{Signal}_{t+1} = \begin{cases} \text{Buy} & \text{if Indicator}_t^{TR} > 0, \\ \text{Sell} & \text{if Indicator}_t^{TR} \leq 0. \end{cases}$$

It is worth emphasising that trading signal Signal_{t+1} is generated at the end of period t and refers to period $t + 1$. If, for example, the trading signal is Buy, this means that a trader buys a financial asset at the period t closing price and holds it over the subsequent period $t + 1$. If the trader owns this asset over period t, he keeps its possession over the subsequent period.

4.2 Momentum Rule

We start with the Momentum (MOM) rule which seemingly has nothing to do with moving averages. However, in the subsequent chapter we show that this rule is inherently related to the rules based on moving averages. The Momentum rule represents the simplest and most basic market timing rule. In this rule, the last closing price P_t is compared with the closing price $n - 1$ periods ago,[2] P_{t-n+1}. A Buy signal is generated when the last closing price is greater than the closing price $n - 1$ periods ago. Implicitly, this rule assumes that if market prices have been increasing (decreasing) over the last $n - 1$ periods, the prices will continue to increase (decrease) over the subsequent period. In other words, the $(n-1)$-period trend will continue in the future. In the scientific literature, the advantages of the Momentum rule are documented by Moskowitz, Ooi, and Pedersen (2012). In the popular literature, the use of this rule is advocated by Antonacci (2014).

Formally, the technical trading indicator for the Momentum rule is computed as

$$\text{Indicator}_t^{\text{MOM}(n)} = MOM_t(n) = P_t - P_{t-n+1}. \tag{4.1}$$

Figure 4.1, bottom panel, plots the values of the technical trading indicator of the $MOM(200)$ rule computed using the daily prices of the S&P 500 index over the period from January 1997 to December 2006. The top panel in this figure plots the values of the index. The shaded areas in this plot indicate the periods where this rule generates a Sell signal.

---

[2]In our notation, n denotes the size of the window used to compute a trading indicator. The most recent price observation in a window is P_t, whereas the most distant price observation in a window is P_{t-n+1}.

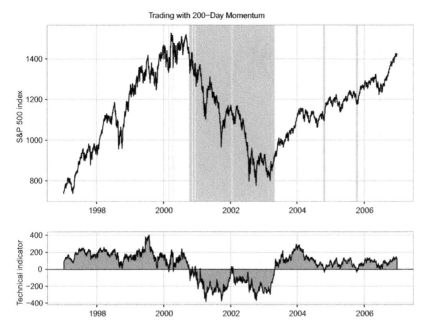

Fig. 4.1 Trading with 200-day Momentum rule. The *top panel* plots the values of the S&P 500 index over the period from January 1997 to December 2006. The *shaded areas* in this plot indicate the periods where this rule generates a Sell signal. The *bottom panel* plots the values of the technical trading indicator

4.3   Moving Average Change of Direction Rule

We proceed to the Moving Average Change of Direction (ΔMA) rule. Even though the use of this rule is not widespread among traders, the idea behind this rule is based on a straightforward principle: if market prices are trending upward (downward), the value of a moving average of prices tends to increase (decrease). In this rule, the most recent value of a moving average is compared with the value of this moving average in the preceding period. A Buy signal is generated when the value of a moving average has increased over the last period.

Formally, the technical trading indicator for the Moving Average Change of Direction rule is computed as

$$\text{Indicator}_t^{\Delta\text{MA}(n)} = MA_t(n) - MA_{t-1}(n). \tag{4.2}$$

Figure 4.2, bottom panel, plots the values of the technical trading indicator of the $\Delta EMA(200)$ rule computed using the daily prices of the S&P 500 index over the period from January 1997 to December 2006. The top panel

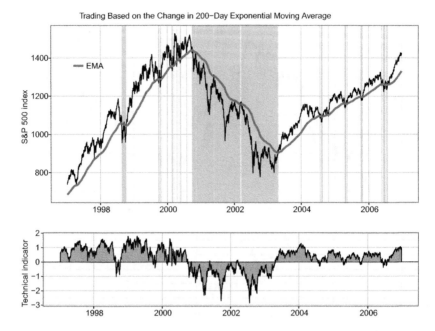

Fig. 4.2 Trading based on the change in 200-day EMA. The *top panel* plots the values of the S&P 500 index over the period from January 1997 to December 2006, as well as the values of EMA(200). The *shaded areas* in this plot indicate the periods where this rule generates a Sell signal. The *bottom panel* plots the values of the technical trading indicator

in this figure plots the values of the index and the 200-day EMA. The shaded areas in this plot indicate the periods where this rule generates a Sell signal.

4.4   Price Minus Moving Average Rule

The Price Minus Moving Average (P-MA) rule is the oldest and one of the most popular trading rules that use moving averages. Gartley (1935) is regarded as the pioneering book where the author laid the foundations for technical trading based on moving averages of prices. In the same book, the author documented the profitability of trading with 200-day SMA. In the scientific literature, the superiority of the 200-day SMA strategy (over the corresponding buy-and-hold strategy) was documented, among others, by Brock, Lakonishok, and LeBaron (1992), Siegel (2002), Okunev and White (2003), Faber (2007), Gwilym, Clare, Seaton, and Thomas (2010), Kilgallen (2012), Clare, Seaton, Smith, and Thomas (2013), and Pätäri and Vilska (2014).

The principle behind this rule is based on the lagging property of a moving average. Specifically, in Chap. 2 we showed explicitly that, when stock prices are trending upward, the moving average lies below the price. In contrast, when stock prices are trending downward, the moving average lies above the price. Therefore, to identify the direction of the trend, in this rule the last closing price is compared with the value of a moving average. A Buy signal is generated when the last closing price is above the moving average. Otherwise, if the last closing price is below the moving average, a Sell signal is generated. Formally, the technical trading indicator for the Price Minus Moving Average rule is computed as

$$\text{Indicator}_t^{\text{P-MA}(n)} = P_t - MA_t(n).$$

Whereas in the Moving Average Change of Direction rule any type of a moving average can be used in principle, in the Price Minus Moving Average rule one needs a moving average that clearly lags behind the time series of prices. Therefore, in this rule, either ordinary moving averages or moving averages of moving averages are used. Typically, traders use SMA in this rule.

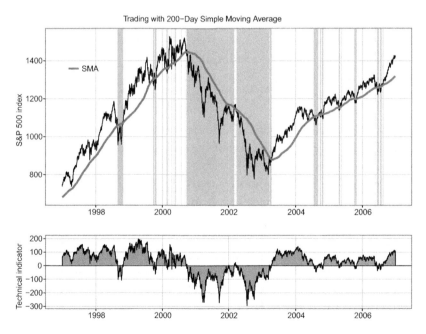

Fig. 4.3 Trading with 200-day Simple Moving Average. The *top panel* plots the values of the S&P 500 index over the period from January 1997 to December 2006, as well as the values of SMA(200). The *shaded areas* in this plot indicate the periods where this rule generates a Sell signal. The *bottom panel* plots the values of the technical trading indicator

Figure 4.3, bottom panel, plots the values of the technical trading indicator of the $P - SMA(200)$ rule computed using the daily prices of the S&P 500 index over the period from January 1997 to December 2006. The top panel in this figure plots the values of the index and the values of the 200-day SMA. The shaded areas in this plot indicate the periods where this rule generates a Sell signal.

4.5   Moving Average Crossover Rule

Most analysts argue that the price is noisy and the Price Minus Moving Average rule produces many false signals. They suggest to address this problem by employing two moving averages in the generation of a trading signal: one shorter average with window size of s and one longer average with window size of $l > s$. This technique is called the Moving Average Crossover (MAC) rule (a.k.a. Double Crossover Method). As a matter of fact, the MAC rule was considered already in Gartley (1935). In this case the technical trading indicator is computed as

$$\text{Indicator}_t^{\text{MAC}(s,l)} = MAC_t(s, l) = MA_t(s) - MA_t(l). \tag{4.3}$$

It is worth noting the obvious relationship

$$\text{Indicator}_t^{\text{MAC}(1,n)} = \text{Indicator}_t^{\text{P-MA}(n)}.$$

In words, the Moving Average Crossover rule reduces to the Price Minus Moving Average rule when the size of the shorter averaging window reduces to one.

A crossover occurs when a shorter moving average crosses either above or below a longer moving average. The former crossover is usually dubbed as a bullish crossover or a "golden cross". The latter crossover is usually dubbed as a bearish crossover or a "death cross". The most typical combination in trading is to use two SMAs with window sizes of 50 and 200 days. Other types of moving averages can also be used in the MAC rule. However, the longer moving average must be of a type that clearly lags behind the price series. A shorter moving average can be of any type including a mixed moving average with less lag time.

Figure 4.4, bottom panel, plots the values of the technical trading indicator of the MAC(50,200) rule computed using the daily prices of the S&P 500 index over the period from January 1997 to December 2006. The top panel in this figure plots the values of the index and the values of 50- and 200-day SMAs.

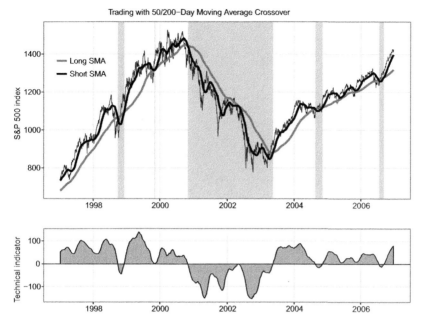

Fig. 4.4 Trading with 50/200-day Moving Average Crossover. The *top panel* plots the values of the S&P 500 index over the period from January 1997 to December 2006, as well as the values of SMA(50) and SMA(200). The *shaded areas* in this plot indicate the periods where this rule generates a Sell signal. The *bottom panel* plots the values of the technical trading indicator

The shaded areas in this plot indicate the periods where this rule generates a Sell signal. It is instructive to compare the number of Sell signals in the P-SMA(200) rule (illustrated in Fig. 4.3) with the number of Sell signals in the MAC(50,200) rule (illustrated in Fig. 4.4). Whereas over the 10-year period 1997–2006 the P-SMA(200) rule generated 40 Sell signals, the MAC(50,200) rule generated only 5 Sell signals. That is, replacing the P-SMA(200) rule with the MAC(50,200) rule produces an impressive 8-fold reduction in transaction costs.[3]

4.6   Using Multiple Moving Averages

If using two moving averages is better than one, then maybe using three or more is even better? Some analysts think so and use multiple moving averages.

---

[3]Note that this 8-fold reduction in transaction costs is achieved only when daily data are used. At a weekly or monthly frequency, the reduction in transaction costs is much lower.

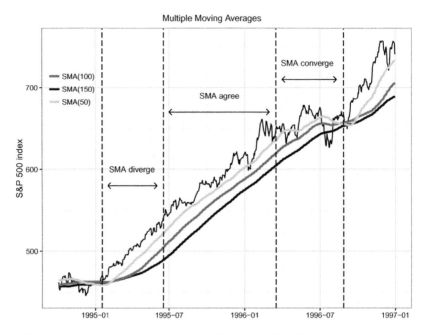

Fig. 4.5 Illustration of a moving average ribbon as well as the common interpretation of the dynamics of multiple moving averages in a ribbon

This technique is often called a moving average ribbon.[4] For the sake of illustration, Fig. 4.5 plots the daily prices of the S&P 500 index over the period from January 1996 to December 1996 as well as a moving average ribbon created using 50-, 100-, and 150-day SMAs. Analysts use ribbons to judge the strength of the trend. Ribbons are also used to identify the trend reversals. The common interpretation of the dynamics of multiple moving averages in a ribbon is as follows:

- When all moving averages are moving in the same direction (that is, parallel), the trend is said to be strong because all of them are largely in agreement.
- When moving averages in a ribbon start to converge or diverge, a trend change has already begun to occur.
- When all moving averages converge and fluctuate more than usual, the price moves sideways.

As a matter of fact, the dynamics of multiple moving averages in a ribbon satisfy the property of moving averages established in Chap. 2. This property

---

[4]See also Guppy (2007) where the author presents his Guppy Multiple Moving Average indicator based on 6 short-term moving averages and 6 long-term moving averages.

says that, when prices trend steadily, all moving averages move parallel in a graph. A change in the direction of the price trend causes moving averages with various average lag times to move in different directions. Therefore, when analysts observe that all moving averages move parallel, this only means that the prices have been steadily trending in the recent past. If, for example, after a period of moving upward in the same direction, the moving averages in a ribbon begin to converge, this means that shorter (and faster) moving averages start to react to a decrease in the price, while longer (and slower) moving averages continue to move upward through inertia.

4.7 Moving Average Convergence/Divergence Rule

A different approach to the generation of trading signals is proposed by Gerald Appel. Specifically, he proposed the Moving Average Convergence/Divergence (MACD) rule which is a combination of three EMAs.[5] The first step in the application of this rule is to compute the regular MAC indicator using two EMAs

$$MAC_t(s, l) = EMA_t(s) - EMA_t(l).$$

Recall that in the regular MAC rule a Buy signal is generated when the shorter moving average is above the longer moving average. In the late 1970s, Gerald Appel suggested to generate a Buy (Sell) signal when MAC increases (decreases). Specifically, in this case a Buy (Sell) signal is generated when the shorter moving average increases (decreases) faster than the longer moving average. In principle, in this approach the generation of a trading signal can be done similarly to that in the Moving Average Change of Direction rule

$$\text{Indicator}_t^{\Delta MAC(s,l)} = MAC_t(s, l) - MAC_{t-1}(s, l).$$

Apparently, Gerald Appel noticed that the ΔMAC rule generates many false signals. In order to reduce the number of false signals, Gerald Appel suggested additionally that a directional movement in MAC must be confirmed by a delayed and smoothed version of MAC. As a result, in the MACD rule the technical trading indicator is computed as

$$\text{Indicator}_t^{MACD(s,l,n)} = MAC_t(s, l) - EMA_t(n, MAC(s, l)). \qquad (4.4)$$

---

[5] For a detailed presentation of the MACD rule, see Appel (2005).

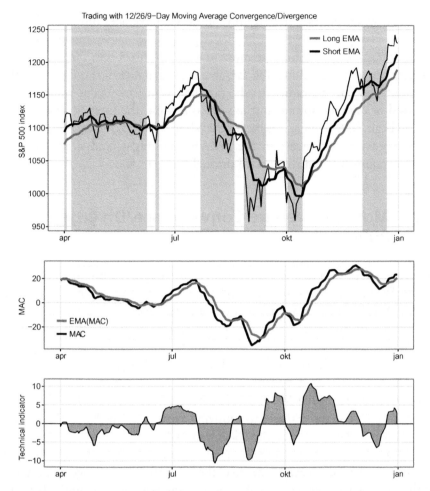

Fig. 4.6 Trading with 12/29/9-day Moving Average Convergence/Divergence rule. The *top panel* plots the values of the S&P 500 index and the values of 12- and 29-day EMAs. The *shaded areas* in this panel indicate the periods where this rule generates a Sell signal. The *middle panel* plots the values of MAC(12,29) and EMA(9,MAC(12,29)). The *bottom panel* plots the values of the technical trading indicator of the MACD(12,29,9) rule

The principle behind the computation of the trading indicator of the MACD rule is the same as that in the Price Minus Moving Average rule. In particular, if MAC is trending upward (downward), a moving average of MAC tends to be below (above) MAC.

Figure 4.6, bottom panel, plots the values of the technical trading indicator of the MACD(12,29,9) rule[6] computed using the daily prices of the S&P 500 index over the period from April 1998 to December 1998. The top panel

---

[6]When traders use the MACD rule, the most popular combination in practice is to use moving averages of 12, 29, and 9 days.

in this figure plots the values of the index as well as the values of 12- and 29-day EMAs. The shaded areas in this plot indicate the periods where this rule generates a Sell signal. The middle panel in this figure plots the values of the MAC(12,29) and the EMA(9,MAC(12,29)).

It is worth noting that the MACD rule, as its name suggests, is devised to generate trading signals when the two moving averages in the MAC indicator either converge or diverge. As a result, the trading signals are generated when the trend either strengthens or weakens. For example, when the price moves upward with an increasing speed, the shorter moving average increases faster than the longer moving average. If the shorter moving average is located above (below) the longer moving average, the two moving averages diverge (converge). Because the value of the MAC increases, the MACD rule generates a Buy signal regardless of the location of the shorter moving average relative to the location of the longer moving average.

Last but not least, it is important to emphasize that, since the MACD rule is devised to react to the changes in the price trend, the MACD rule is most suited when the price trend often changes its direction. In contrast, when prices trend steadily, both the moving averages move parallel. In this case even small changes in the price dynamics are able to generate lots of false trading signals.

4.8   Limitations of Moving Average Trading Rules

When prices trend steadily upward or downward, moving averages easily identify the direction of the trend. In these cases, all moving average trading strategies generate correct Buy and Sell trading signals, albeit with some delay. However, as it was observed already in Gartley (1935), when prices trend sideways, moving average strategies tend to generate many false signals, otherwise known as "whipsaws". The reason for these whipsaw trades are considered in Chap. 2. Specifically, in this chapter we showed explicitly that, when stock prices trend sideways, the value of a moving average is close to the last closing price. As a result, even small fluctuations in the price may result in a series of unnecessary trades.

To demonstrate this issue, an illustration is provided in Fig. 4.7. In particular, this figure plots the daily prices of the S&P 500 index over the period from July 1999 to October 2000 as well as the values of 200-day SMA. During this period, that lasted 15 months, the P-SMA(200) rule generated 13 Sell signals. All of them were quickly reversed and, therefore, these Sell signals did not work out and resulted in a series of small losses.

Fig. 4.7 Trading with 200-day Simple Moving Average. The figure plots the values of the S&P 500 index over the period from July 1999 to October 2000, as well as the values of SMA(200). The shaded areas in this plot indicate the periods where this rule generates a Sell signal

There are several remedies that allow a trader to reduce the number of whipsaw trades. One possibility is to use the MAC rule. For example, over the same period as that in Fig. 4.7, the MAC(50,200) rule did not generate a single Sell signal. The other possibility to reduce the number of whipsaw trades is to use a Moving Average Envelope. Specifically, a moving average envelope consists of two boundaries above and below a moving average. The distance from the moving average and a boundary of the envelope is usually specified as a percentage (for example, 1%). As long as the price lies within these two boundaries, no trading takes place. A Buy (Sell) signal is generated when the price crosses the upper (lower) boundary of the envelope. Formally, denote by $MA_t(n)$ the moving average of prices over a window of size n and by p the envelope percentage. The upper and lower boundaries of the moving average envelope are computed by

$$L_t = MA_t(n) \times (1 - p), \quad U_t = MA_t(n) \times (1 + p).$$

Mathematically, the trading signal is generated according to:

$$\text{Signal}_{t+1} = \begin{cases} \text{Buy} & \text{if } P_t > U_t, \\ \text{Sell} & \text{if } P_t < L_t, \\ \text{Signal}_t & \text{if } L_t \leq P_t \leq U_t. \end{cases}$$

Notice that, when the price lies within the two boundaries, the trading signal for the period $t + 1$ equals the trading signal for the previous period t. For example, when the price crosses (from below to above) the upper boundary, a Buy signal is generated. The trading signal remains Buy until the price crosses (from above to below) the lower boundary.

The other serious limitation of all moving average trading rules arises from the lagging nature of a moving average. Specifically, a turning point in a trend is always recognized with some delay. Therefore, in order for a trend following strategy to generate profitable trading signals, the duration of a trend should be long enough.

To illustrate this point, Fig. 4.8 plots the daily prices of the S&P 500 index over the whole year of 1998, as well as the values of 50- and 200-day SMAs.

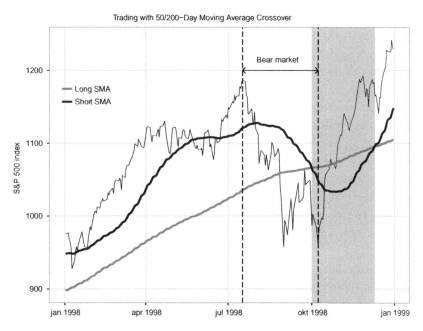

Fig. 4.8 Trading with 50/200-day Moving Average Crossover. The figure plots the values of the S&P 500 index over the period from January 1998 to December 1998, as well as the values of 50- and 200-day SMAs. The *shaded area* in this plot indicates the period where this rule generates a Sell signal

This particular period covers the 1998 Russian financial crisis and the US stock market reaction to this crisis. Specifically, the S&P 500 index reacted to this crisis beginning from 18 of July 1998. That is, 17 of July 1998 was a divider between a bullish and a bearish trend. However, the bear market that began on 18 of July lasted for less than 3 months. The S&P 500 index started to recover beginning from 8 of October 1998. Yet, the most popular among practitioners MAC(50,200) rule generated a Sell signal only on 2 of October 1998, at the end of the bear market. The subsequent Buy signal was generated on 11 of December 1998. Thus, because both Sell and Buy signals were generated with a delay, virtually this whole "Sell" period overlapped with the bull market. The traders who used this rule suffered heavy losses because they were forced to "sell low and buy high".

Unfortunately, no remedy exists to deal with the lagging property of a moving average. It is worth noting that all techniques that reduce the number of whipsaw trades usually achieve that at the expense of increasing the delay in turning point identification.

4.9   Chapter Summary

The success of a trend following strategy depends on its ability to timely identify the direction of the trend in prices. However, fluctuations in prices make it difficult to recognize the direction of the price trend. Moving averages are often used to smooth these fluctuations in order to highlight the underlying trend.

Even though the concept of trend following is simple ("jump on a trend and ride it"), there is no unique practical realization of a trend following strategy. There are trend following rules that do not employ moving averages. There are trend following rules that use only one moving average. But even in this case, there are two possible methods of generation of a Buy signal: either when the value of a moving average increases, or when the value of a moving average lies below the price. In addition, there are trend following rules that employ two, three, and even multiple moving averages. As a rule, the supplementary moving averages are used to improve the performance of a moving average trading strategy.

The moving average trading strategies are advantageous when the trend is strong and long-lasting. However, the advantages of the moving average strategies may disappear completely when the trend is weak. Due to the lagging nature of any moving average, the advantages of the moving average strategies may disappear even if the trend is strong but short-lasting.

References

Antonacci, G. (2014). *Dual momentum investing: An innovative strategy for higher returns with lower risk.* McGraw-Hill Education.

Appel, G. (2005). *Technical analysis: Power tools for active investors.* FT Prentice Hall.

Brock, W., Lakonishok, J., & LeBaron, B. (1992). Simple technical trading rules and the stochastic properties of stock returns. *Journal of Finance, 47*(5), 1731–1764.

Clare, A., Seaton, J., Smith, P. N., & Thomas, S. (2013). Breaking into the blackbox: Trend following, stop losses and the frequency of trading—the case of the S&P500. *Journal of Asset Management, 14*(3), 182–194.

Faber, M. T. (2007). A quantitative approach to tactical asset allocation. *Journal of Wealth Management, 9*(4), 69–79.

Gartley, H. M. (1935). *Profits in the stock market.* Lambert Gann Pub.

Guppy, D. (2007). *Trend trading: A seven step approach to success.* Wrightbooks.

Gwilym, O., Clare, A., Seaton, J., & Thomas, S. (2010). Price and momentum as robust tactical approaches to global equity investing. *Journal of Investing, 19*(3), 80–91.

Kilgallen, T. (2012). Testing the simple moving average across commodities, global stock indices, and currencies. *Journal of Wealth Management, 15*(1), 82–100.

Moskowitz, T. J., Ooi, Y. H., & Pedersen, L. H. (2012). Time series momentum. *Journal of Financial Economics, 104*(2), 228–250.

Okunev, J., & White, D. (2003). Do momentum-based strategies still work in foreign currency markets? *Journal of Financial and Quantitative Analysis, 38*(2), 425–447.

Pätäri, E., & Vilska, M. (2014). Performance of moving average trading strategies over varying stock market conditions: The finnish evidence. *Applied Economics, 46*(24), 2851–2872.

Siegel, J. (2002). *Stocks for the long run.* McGraw-Hill Companies.

5

Anatomy of Trading Rules

5.1 Preliminaries

In our context, a technical trading indicator can be considered as a combination of a specific technical trading rule with a particular moving average of prices. In the preceding chapters of this book we show that there are many technical trading rules, as well as there are many popular types of moving averages. As a result, there exist a vast number of potential trading indicators based on moving averages of prices. So far, the development in this field has consisted in proposing new ad-hoc trading rules and using more elaborate types of moving averages in the existing rules. Each new proposed rule (or moving average) appears on the surface as something unique. Often this new proposed rule (or moving average) is said to be better than its competitors; such a claim is usually supported by colorful narratives and anecdotal evidence.

The existing situation in the field of market timing with moving averages is as follows. Technical traders are overwhelmed by the variety of choices between different trading indicators. Because traders do not really understand the response characteristics of the trading indicators they use, the selection of a trading indicator is made based mainly on intuition rather than any deeper analysis of commonalities and differences between miscellaneous choices for trading rules and moving averages. It would be no exaggeration to say that the existing situation resembles total chaos and mess from the perspective of a newcomer to this field.

The ultimate goal of this chapter is to bring some order to the chaos in the field of market timing with moving averages. We offer a framework that can be used to uncover the anatomy of market timing rules with moving averages of prices. Specifically, we present a methodology for examining how the value

© The Author(s) 2017
V. Zakamulin, *Market Timing with Moving Averages*, New Developments
in Quantitative Trading and Investment, DOI 10.1007/978-3-319-60970-6_5

of a trading indicator is computed. Then using this methodology we study the computation of trading indicators in many market timing rules and analyze the commonalities and differences between the rules. Our analysis gives a new look to old indicators and offers a new and very insightful re-interpretation of the existing market timing rules.

To begin with, as motivation, consider the following example. It has been known for years that there is a relationship between the Momentum rule and the Simple Moving Average Change of Direction rule.[1] In particular, note that

$$SMA_t(n-1) - SMA_{t-1}(n-1) = \frac{P_t - P_{t-n+1}}{n-1} = \frac{MOM_t(n)}{n-1}. \quad (5.1)$$

Therefore

$$\text{Indicator}_t^{\Delta SMA(n-1)} \equiv \text{Indicator}_t^{MOM(n)}, \quad (5.2)$$

where the mathematical symbol "$\equiv$" means "equivalence". The equivalence of two technical indicators stems from the following property: *the multiplication of a technical indicator by any positive real number produces an equivalent technical indicator.* This is because the trading signal is generated depending on the sign of the technical indicator. The formal presentation of this property:

$$\text{sgn}\left(a \times \text{Indicator}_t^{TR}\right) = \text{sgn}\left(\text{Indicator}_t^{TR}\right), \quad (5.3)$$

where a is any positive real number and $\text{sgn}(\cdot)$ is the mathematical *sign function* defined by

$$\text{sgn}(x) = \begin{cases} 1 & \text{if } x>0, \\ 0 & \text{if } x=0, \\ -1 & \text{if } x<0. \end{cases}$$

To see the validity of relation (5.2), observe from Eq. (5.1) that if $SMA_t(n-1) - SMA_{t-1}(n-1) > 0$ then $MOM_t(n) > 0$ and vice versa. In other words, the Simple Moving Average Change of Direction rule, $\Delta SMA(n-1)$, generates a Buy (Sell) trading signal when the Momentum rule, $MOM_t(n)$, generates a Buy (Sell) trading signal.

What else can we say about the relationship between different market timing rules? Are there other seemingly different rules that generate similar trading signals? Which rules differ only a little and which rules differ substantially?

---

[1] See, for example, http://en.wikipedia.org/wiki/Momentum_(technical_analysis).

This chapter offers answers to these questions and demonstrates that all market timing rules considered in this book are closely interconnected. In particular, we are going to show that the computation of a technical trading indicator for every market timing rule, based on either one or multiple moving averages, can be interpreted as the computation of a single weighted moving average of price changes over the averaging window. More formally, we will demonstrate that the computation of a technical trading indicator for every market timing rule can be written as

$$\text{Indicator}_t^{TR(n)} = \sum_{i=1}^{n-1} \pi_i \Delta P_{t-i}, \tag{5.4}$$

where, recall, $\Delta P_{t-i} = P_{t-i+1} - P_{t-i}$ denotes the price change and π_i is the weight of the price change ΔP_{t-i} in the computation of a weighted moving average of price changes. Therefore, despite a great variety of trading indicators that are computed seemingly differently at the first sight, the only real difference between the diverse trading indicators lies in the weighting function used to compute the moving average of price changes. In addition, we will show that the weights π_i can be normalized for majority of market timing rules.

5.2 Momentum Rule

The computation of the technical trading indicator for the Momentum rule can equivalently be written as

$$\text{Indicator}_t^{MOM(n)} = MOM_t(n) = P_t - P_{t-n+1}$$

$$= (P_t - P_{t-1}) + (P_{t-1} - P_{t-2}) + \ldots + (P_{t-n+2} - P_{t-n+1}) = \sum_{i=1}^{n-1} \Delta P_{t-i}. \tag{5.5}$$

Consequently, using property (5.3), the computation of the technical indicator for the Momentum rule is equivalent[2] to the computation of the equally

---

[2]When we apply property (5.3), we use $a = \frac{1}{n-1}$. In virtually all cases the value of a is chosen to normalize the set of weights. In this particular case a also equals the weight of each price change in the computation of the moving average of price change.

weighted moving average of price changes (in a window which contains n consequent prices):

$$\text{Indicator}_t^{\text{MOM}(n)} \equiv \frac{1}{n-1} \sum_{i=1}^{n-1} \Delta P_{t-i}. \tag{5.6}$$

Written in this form, it becomes evident that the Momentum rule can also be classified as a moving average trading rule. The important distinction is that this rule is based on a moving average of price changes, not prices.

5.3   Price Minus Moving Average Rule

First, we derive the relationship between the Price Minus Moving Average rule and the Momentum rule:

$$
\begin{aligned}
\text{Indicator}_t^{\text{P-MA}(n)} &= P_t - MA_t(n) = P_t - \frac{\sum_{i=0}^{n-1} w_i P_{t-i}}{\sum_{i=0}^{n-1} w_i} = \frac{\sum_{i=0}^{n-1} w_i P_t - \sum_{i=0}^{n-1} w_i P_{t-i}}{\sum_{i=0}^{n-1} w_i} \\
&= \frac{\sum_{i=1}^{n-1} w_i (P_t - P_{t-i})}{\sum_{i=0}^{n-1} w_i} = \frac{\sum_{i=1}^{n-1} w_i MOM_t(i+1)}{\sum_{i=0}^{n-1} w_i}.
\end{aligned} \tag{5.7}
$$

Observe that weight w_0 is absent in the numerator of the last fraction above. Therefore the sum of the weights in the numerator is not equal to the sum of the weights in the denominator. However, using property (5.3), we can delete weight w_0 from the denominator. As a result, the relation above can be conveniently re-written as

$$\text{Indicator}_t^{\text{P-MA}(n)} \equiv \frac{\sum_{i=1}^{n-1} w_i MOM_t(i+1)}{\sum_{i=1}^{n-1} w_i}. \tag{5.8}$$

In this form, the derived equivalence relation says that the computation of the technical trading indicator for the Price Minus Moving Average rule, $P_t - MA_t(n)$, is equivalent to the computation of the weighted moving average of technical indicators for the Momentum rules, $MOM_t(i+1)$, for $i \in [1, n-1]$. It is worth noting that the weighting function for computing the moving average of the Momentum technical indicators is virtually the same as the weighting function for computing the weighted moving average $MA_t(n)$.

Second, we use identity (5.5) and rewrite the numerator of the last fraction in (5.7) as a double sum

$$\sum_{i=1}^{n-1} w_i MOM_t(i+1) = \sum_{i=1}^{n-1} w_i \sum_{j=1}^{i} \Delta P_{t-j}.$$

Interchanging the order of summation in the double sum above yields

$$\sum_{i=1}^{n-1} w_i \sum_{j=1}^{i} \Delta P_{t-j} = \sum_{j=1}^{n-1} \left(\sum_{i=j}^{n-1} w_i \right) \Delta P_{t-j}. \qquad (5.9)$$

This result tells us that the numerator of the last fraction in (5.7) is a weighted sum of the price changes over the averaging window, where the weight of ΔP_{t-j} equals $\sum_{i=j}^{n-1} w_i$. Thus, another alternative expression for the computation of the technical indicator for the Price Minus Moving Average rule is given by

$$\text{Indicator}_t^{\text{P-MA}(n)} = \frac{\sum_{j=1}^{n-1} \left(\sum_{i=j}^{n-1} w_i \right) \Delta P_{t-j}}{\sum_{i=0}^{n-1} w_i} = \sum_{j=1}^{n-1} \phi_j \Delta P_{t-j}, \qquad (5.10)$$

where ϕ_j is given by Eq. (2.6). It is worth noting that we could derive this result much more easily using Eq. (2.7) for the alternative representation of a weighted moving average. However, a longer two-step derivation allows us to show that the computation of the technical trading indicator for the Price Minus Moving Average rule can equivalently be interpreted in two alternative ways: as a computation of the weighted moving average of Momentum rules, and as a computation of the weighted moving average of price changes.

In the same manner as in Sect. 2.3, we can analyse the properties of the technical trading indicator for the Price Minus Moving Average rule by assuming that the price change follows a Random Walk process with a drift: $\Delta P_{t-j} = E[\Delta P] + \sigma \varepsilon_j$. In this case the expected value of the technical indicator is given by

$$E[P_t - MA_t(n)] = \sum_{j=1}^{n-1} \phi_j E[\Delta P] = \text{Lag time}(MA) \times E[\Delta P]. \qquad (5.11)$$

If the prices trend upward ($E[\Delta P] > 0$), in order a Buy signal is generated, the expected value of this trading indicator should be positive. This requires that the average lag time of a moving average must be strictly positive (Lag time(MA) > 0). In other words, this rule requires a moving average that clearly lags behind the price trend. Otherwise, if a moving average in this rule has zero lag time, this trading indicator becomes a random noise generator.[3] In addition, Eq. (5.11) implies that when the trend is sideways (meaning that $E[\Delta P] = 0$), then again this trading indicator becomes a random noise generator.

Yet another property of the technical trading indicator for the Price Minus Moving Average rule appears due to the method of computation of weights ϕ_j. Since

$$\phi_j = \frac{\sum_{i=j}^{n-1} w_i}{\sum_{i=0}^{n-1} w_i},$$

in case all weights w_i are strictly positive, the sequence of weights ϕ_j is decreasing with increasing j

$$\phi_1 > \phi_2 > \ldots > \phi_{n-1}.$$

Consequently, in this case, regardless of the shape of the weighting function for prices w_i, the weighting function ϕ_j always over-weights the most recent price changes. Specifically, if, for example, all prices are equally weighted in a moving average, then the application of the Price Minus Moving Average rule leads to overweighting the most recent price changes. If the price weighting function of a moving average is already designed to overweight the most recent prices, then generally the trading signal in this rule is computed with a much stronger overweighting the most recent price changes. Probably the only exception is the Price Minus Exponential Moving Average rule; this will be demonstrated below.

Before going further, observe that the weights in Eq. (5.10) are not normalized. This issue can be easily fixed by using property (5.3) and rewriting Eq. (5.10) as

$$\text{Indicator}_t^{\text{P-MA}(n)} \equiv \frac{\sum_{j=1}^{n-1} \upsilon_j \Delta P_{t-j}}{\sum_{j=1}^{n-1} \upsilon_j}, \quad \text{where } \upsilon_j = \sum_{i=j}^{n-1} w_i. \quad (5.12)$$

---

[3]This is because in this case the expected value of the trading indicator equals zero. Therefore the value of the difference $P_t - MA_t(n)$ is related to the weighted sum of random disturbances $\sigma \varepsilon_j$. This sum is also a random variable with zero mean.

Let us now, on the basis of (5.12), derive the closed-form expressions for the computation of technical indicator of the Price Minus Moving Average rule for all ordinary moving averages considered in Sect. 3.1. We start with the Simple Moving Average which is the equally weighted moving average of prices. In this case the weight of ΔP_{t-j} is given by

$$v_j = \sum_{i=j}^{n-1} w_i = \sum_{i=j}^{n-1} 1 = n - j. \tag{5.13}$$

Consequently, the equivalent representation for the computation of the technical indicator for the Price Minus Simple Moving Average rule is given by

$$\text{Indicator}_t^{\text{P-SMA}(n)} \equiv \frac{\sum_{j=1}^{n-1}(n-j)\Delta P_{t-j}}{\sum_{j=1}^{n-1}(n-j)} = \frac{(n-1)\Delta P_{t-1} + (n-2)\Delta P_{t-2} + \ldots + \Delta P_{t-n+1}}{(n-1) + (n-2) + \ldots + 1}. \tag{5.14}$$

The resulting formula suggests that alternatively we can interpret the computation of the technical indicator for the Price Minus Simple Moving Average rule as the computation of a Linearly Weighted Moving Average of price changes.

We next consider the Linear Moving Average. In this case the weight of ΔP_{t-j} is given by

$$v_j = \sum_{i=j}^{n-1} w_i = \sum_{i=j}^{n-1}(n-j) = \frac{(n-j)(n-j+1)}{2}, \tag{5.15}$$

which is the sum of the terms of arithmetic sequence from 1 to $n - j$ with the common difference of 1. As the result, the equivalent representation for the computation of the technical indicator for the Price Minus Linear Moving Average rule is given by

$$\text{Indicator}_t^{\text{P-LMA}(n)} \equiv \frac{\sum_{j=1}^{n-1} \frac{(n-j)(n-j+1)}{2} \Delta P_{t-j}}{\sum_{j=1}^{n-1} \frac{(n-j)(n-j+1)}{2}}. \tag{5.16}$$

Finally we consider the (infinite) Exponential Moving Average which is computed as

$$EMA_t(\lambda) = \frac{P_t + \lambda P_{t-1} + \lambda^2 P_{t-2} + \lambda^3 P_{t-3} + \dots}{1 + \lambda + \lambda^2 + \lambda^3 + \dots}.$$

In this case the size of the averaging window $n \to \infty$ and the weight of ΔP_{t-i} is given by

$$v_j = \sum_{i=j}^{\infty} w_i = \sum_{i=j}^{\infty} \lambda^i = \frac{\lambda^j}{1 - \lambda}, \tag{5.17}$$

which is the sum of the terms of a geometric sequence with the initial term λ^j and the common ratio λ. Consequently, the equivalent presentation for the computation of the technical indicator for the Price Minus Exponential Moving Average rule is given by

$$\text{Indicator}_t^{\text{P-EMA}(\lambda)} \equiv \frac{\sum_{j=1}^{\infty} \lambda^j \Delta P_{t-j}}{\sum_{j=1}^{\infty} \lambda^j} = (1 - \lambda) \sum_{j=1}^{\infty} \lambda^{j-1} \Delta P_{t-j}. \tag{5.18}$$

In words, the computation of the trading indicator for the Price Minus Exponential Moving Average rule is equivalent to the computation of the Exponential Moving Average of price changes. It is worth noting that this is probably the only trading indicator where the weighting function for the computation of moving average of prices is identical to the weighting function for the computation of moving average of price changes.

We remind the reader that instead of notation $EMA_t(\lambda)$ one uses notation $EMA_t(n)$ where n denotes the size of the averaging window in a Simple Moving Average with the same average lag time as in $EMA_t(\lambda)$. The value of the decay factor in $EMA_t(n)$ is computed as $\lambda = \frac{n-1}{n+1}$.

For the sake of illustration, Fig. 5.1 plots the shapes of the price change weighting functions in the Momentum (MOM) rule and four Price Minus Moving Average rules: Price Minus Simple Moving Average (P-SMA) rule, Price Minus Linear Moving Average (P-LMA) rule, Price Minus Exponential Moving Average (P-EMA) rule, and Price Minus Triangular Moving Average (P-TMA) rule. In all rules, the size of the averaging window equals $n = 30$. Observe that in all but the Momentum rule the weighting function overweights the most recent price changes.

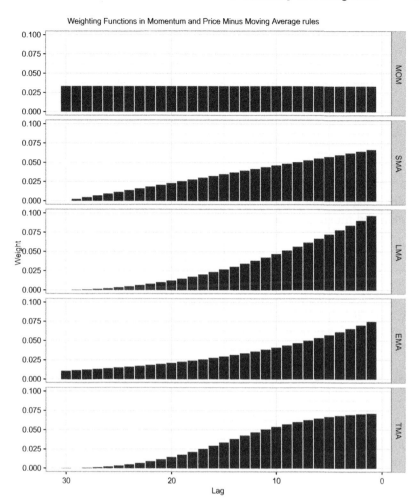

Fig. 5.1 The shapes of the price change weighting functions in the Momentum (MOM) rule and four Price Minus Moving Average rules: Price Minus Simple Moving Average (P-SMA) rule, Price Minus Linear Moving Average (P-LMA) rule, Price Minus Exponential Moving Average (P-EMA) rule, and Price Minus Triangular Moving Average (P-TMA) rule. In all rules, the size of the averaging window equals $n = 30$. The weights of the price changes in the P-EMA rule are cut off at lag 30

5.4 Moving Average Change of Direction Rule

The value of this technical trading indicator is based on the difference of two weighted moving averages computed at times t and $t - 1$ respectively. We assume that in each moving average the size of the averaging window equals

$n - 1$. The reason for this assumption is to ensure that the trading indicator is computed over the window of size n. The straightforward computation yields

$$
\begin{aligned}
\text{Indicator}_t^{\Delta MA(n-1)} &= MA_t(n-1) - MA_{t-1}(n-1) = \frac{\sum_{i=0}^{n-2} w_i P_{t-i}}{\sum_{i=0}^{n-2} w_i} - \frac{\sum_{i=0}^{n-2} w_i P_{t-i-1}}{\sum_{i=0}^{n-2} w_i} \\
&= \frac{\sum_{i=0}^{n-2} w_i (P_{t-i} - P_{t-i-1})}{\sum_{i=0}^{n-2} w_i} = \frac{\sum_{i=0}^{n-2} w_i \Delta P_{t-i-1}}{\sum_{i=0}^{n-2} w_i} = \frac{\sum_{i=1}^{n-1} w_{i-1} \Delta P_{t-i}}{\sum_{i=1}^{n-1} w_{i-1}}.
\end{aligned}
\tag{5.19}
$$

Consequently, the computation of the technical indicator for the Moving Average Change of Direction rule can be interpreted as the computation of the weighted moving average of price changes:

$$
\text{Indicator}_t^{\Delta MA(n-1)} = \frac{\sum_{i=1}^{n-1} w_{i-1} \Delta P_{t-i}}{\sum_{i=1}^{n-1} w_{i-1}}.
\tag{5.20}
$$

From (5.20) we can easily recover the relationship for the case of the Simple Moving Average where $w_{i-1} = 1$ for all i:

$$
\text{Indicator}_t^{\Delta SMA(n-1)} = \frac{\sum_{i=1}^{n-1} \Delta P_{t-i}}{\sum_{i=1}^{n-1} 1} = \frac{1}{n-1} \sum_{i=1}^{n-1} \Delta P_{t-i} \equiv \text{Indicator}_t^{MOM(n)},
\tag{5.21}
$$

where the last equivalence follows from (5.6).

In the case of the Linear Moving Average where $w_{i-1} = n - i$, we derive a new relationship:

$$
\text{Indicator}_t^{\Delta LMA(n-1)} \equiv \frac{\sum_{i=1}^{n-1} (n-i) \Delta P_{t-i}}{\sum_{i=1}^{n-1} (n-i)} \equiv \text{Indicator}_t^{\text{P-SMA}}(n),
\tag{5.22}
$$

where the last equivalence follows from (5.14). Putting it into words, the Price Minus Simple Moving Average rule, $P_t - SMA_t(n)$, prescribes investing in the stocks (moving to cash) when the Linear Moving Average of prices, $LMA_t(n-1)$, increases (decreases).

In the case of the Exponential Moving Average, the resulting expression for the Change of Direction rule can be written as

$$
\text{Indicator}_t^{\Delta EMA(\lambda)} = \frac{\sum_{i=1}^{\infty} \lambda^{i-1} \Delta P_{t-i}}{\sum_{i=1}^{\infty} \lambda^{i-1}} = (1-\lambda) \sum_{j=1}^{\infty} \lambda^{j-1} \Delta P_{t-j}.
\tag{5.23}
$$

Consequently, the computation of the technical indicator for the Exponential Moving Average Change of Direction rule is equivalent to the computation of the (infinite) Exponential Moving Average of price changes. Observe also the similarity between Eqs. (5.23) and (5.18). This similarity implies that the Exponential Moving Average Change of Direction rule is equivalent to the Price Minus Exponential Moving Average rule.

For the sake of illustration, Fig. 5.2 plots the shapes of the price change weighting functions in five Moving Average Change of Direction rules: Simple Moving Average (SMA) Change of Direction rule, Linear (LMA) Moving Average Change of Direction rule, Exponential Moving Average (EMA) Change of Direction rule, Double Exponential Moving Average (EMA(EMA)) Change of Direction rule, and Triangular (TMA) Moving Average Change of Direction rule. In all rules, the size of the averaging window equals $n = 30$.

Finally it is worth commenting that the traders had long ago taken notice of the fact that often a trading signal (Buy or Sell) is generated first by the Price Minus Moving Average rule, then with some delay the same trading signal is generated by the corresponding Moving Average Change of Direction rule. Therefore the traders, who use the Price Minus Moving Average rule, often wait to see whether a trading signal of the Price Minus Moving Average rule is "confirmed" by a trading signal of the corresponding Moving Average Change of Direction rule (see Murphy 1999, Chap. 9). Our analysis provides a simple explanation for the existence of a natural delay between the signals generated by these two rules. Specifically, the delay naturally occurs because the Price Minus Moving Average rule overweights more heavily the most recent price changes than the Moving Average Change of Direction rule computed using the same weighting scheme. Therefore the Price Minus Moving Average rule reacts more quickly to the recent trend changes than the Moving Average Change of Direction rule.

To elaborate on the aforesaid in more details, suppose that the trader uses the Price Minus Simple Moving Average rule and acts only when the signal generated by this rule is confirmed by a corresponding signal generated by the Simple Moving Average Change of Direction rule. Our result (5.22) says that the Price Minus Simple Moving Average rule is equivalent to the Linear Moving Average Change of Direction rule. Consequently, the trader's strategy can equivalently be interpreted as follows: observe the signal generated by the Linear Moving Average Change of Direction rule and wait for the corresponding signal generated by the Simple Moving Average Change of Direction rule. We know from Chap. 3 that the Linear Moving Average has a shorter average lag time than the Simple Moving Average. Therefore, the Linear Moving Average reacts faster to the changes in the direction of the price trend than the Simple

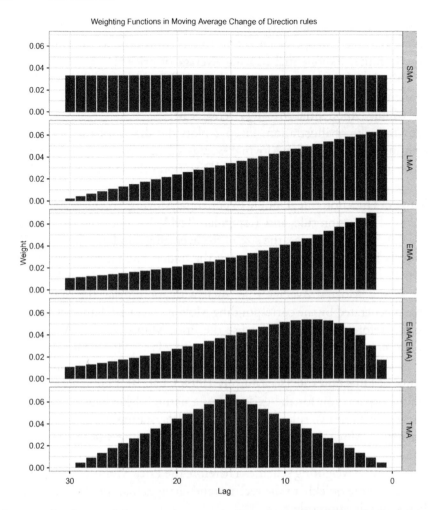

Fig. 5.2 The shapes of the price change weighting functions in five Moving Average Change of Direction rules: Simple Moving Average (SMA) Change of Direction rule, Linear (LMA) Moving Average Change of Direction rule, Exponential Moving Average (EMA) Change of Direction rule, Double Exponential Moving Average (EMA(EMA)) Change of Direction rule, and Triangular (TMA) Moving Average Change of Direction rule. In all rules, the size of the averaging window equals $n = 30$. The weights of the price changes in the $\triangle$EMA and $\triangle$EMA(EMA) rules are cut off at lag 30

Moving Average. As a result, when the trading signal of the Simple Moving Average Change of Direction rule "confirms" the trading signal of the Linear Moving Average Change of Direction rule, it only means that, after a recent break in trend identified by the Linear Moving Average Change of Direction rule, the prices continued to trend in the same direction for a while.

5.5 Moving Average Crossover Rule

The relationship between the Moving Average Crossover rule and the Momentum rule is as follows (here we use the result given by Eq. (5.7))

$$
\begin{aligned}
\text{Indicator}_t^{MAC(s,l)} &= MA_t(s) - MA_t(l) = (P_t - MA_t(l)) - (P_t - MA_t(s)) \\
&= \frac{\sum_{i=1}^{l-1} w_i^l MOM_t(i+1)}{\sum_{i=0}^{l-1} w_i^l} - \frac{\sum_{i=1}^{s-1} w_i^s MOM_t(i+1)}{\sum_{i=0}^{s-1} w_i^s} \\
&= \sum_{i=1}^{l-1} \phi_i^l MOM_t(i+1) - \sum_{i=1}^{s-1} \phi_i^s MOM_t(i+1).
\end{aligned}
\tag{5.24}
$$

Different superscripts in the weights mean that for the same subscript the weights are generally not equal. For example, in case of the Linear Moving Average, $w_i^l = l - i$ whereas $w_i^s = s - i$.

The application of the result given by Eq. (5.10) yields

$$
\begin{aligned}
\text{Indicator}_t^{MAC(s,l)} &= \frac{\sum_{j=1}^{l-1} \left(\sum_{i=j}^{l-1} w_i^l \right) \Delta P_{t-j}}{\sum_{i=0}^{l-1} w_i^l} - \frac{\sum_{j=1}^{s-1} \left(\sum_{i=j}^{s-1} w_i^s \right) \Delta P_{t-j}}{\sum_{i=0}^{s-1} w_i^s} \\
&= \sum_{j=1}^{l-1} \phi_j^l \Delta P_{t-j} - \sum_{j=1}^{s-1} \phi_j^s \Delta P_{t-j}.
\end{aligned}
\tag{5.25}
$$

Therefore the computation of the trading indicator in the Moving Average Crossover rule can be presented as

$$
\text{Indicator}_t^{MAC(s,l)} = \sum_{j=1}^{s-1} \left(\phi_j^l - \phi_j^s \right) \Delta P_{t-j} + \sum_{j=s}^{l-1} \phi_j^l \Delta P_{t-j}.
\tag{5.26}
$$

The computation of the trading indicator in the Moving Average Crossover rule is basically similar to the computation of the trading indicator in the Price Minus Moving Average rule; the only difference is that the shorter moving average is used instead of the last closing price. To understand the effect of using the shorter moving average instead of the last price, we present the computation of the trading indicator in the Price Minus Moving Average rule in the following form (assuming that $l = n$)

$$
\text{Indicator}_t^{P\text{-}MA(l)} = \sum_{j=1}^{l-1} \phi_j^l \Delta P_{t-j} = \sum_{j=1}^{s-1} \phi_j^l \Delta P_{t-j} + \sum_{j=s}^{l-1} \phi_j^l \Delta P_{t-j}.
\tag{5.27}
$$

The comparison of Eqs. (5.26) and (5.27) reveals that the price change weighting functions for both the rules, $MAC(s, l)$ and $P - MA(l)$, are identical beginning from lag s and beyond. In contrast, as compared to the price change weighting function of $P - MA(l)$ rule, the price change weighting function of $MAC(s, l)$ rule assigns smaller weights to the most recent price changes (from lag 1 to lag $s - 1$). Since most typically the price change weighting function in the $P - MA(l)$ rule overweights the most recent price changes, the reduction of weights of the most recent price changes in the $MAC(s, l)$ rule makes its price change weighting function to underweight both the most recent and the most distant price changes.

When the Simple Moving Average is used in both the shorter and longer moving averages, the computation of the trading indicator is given by (see the subsequent appendix for the details of the derivation)

$$\text{Indicator}_t^{SMAC(s,l)} = SMA_t(s) - SMA_t(l) = \sum_{j=1}^{s-1} \frac{(l - s)j}{l \times s} \Delta P_{t-j} + \sum_{j=s}^{l-1} \frac{(l - j)}{l} \Delta P_{t-j}.$$

(5.28)

When the lag number j increases, the price change weighting function in this rule linearly increases till lag s where it attains its maximum. Afterwards, the price change weighting function linearly decreases toward zero.

When the Exponential Moving Average is used in both the shorter and longer moving averages, the computation of the trading indicator is given by (see the subsequent appendix for the details of the derivation)

$$\text{Indicator}_t^{EMAC(s,l)} = EMA_t(\lambda_s) - EMA_t(\lambda_l) = \sum_{j=1}^{\infty} \left(\lambda_l^j - \lambda_s^j \right) \Delta P_{t-j},$$

(5.29)

where

$$\lambda_l = \frac{l - 1}{l + 1}, \quad \lambda_s = \frac{s - 1}{s + 1}.$$

Again, when the lag number j increases, the price change weighting function first increases, attains the maximum, then decreases toward zero. Specifically, the price change weighting function attains its maximum at lag

$$j = \frac{\ln \left(\frac{\ln(\lambda_s)}{\ln(\lambda_l)} \right)}{\ln \left(\frac{\lambda_l}{\lambda_s} \right)}.$$

(5.30)

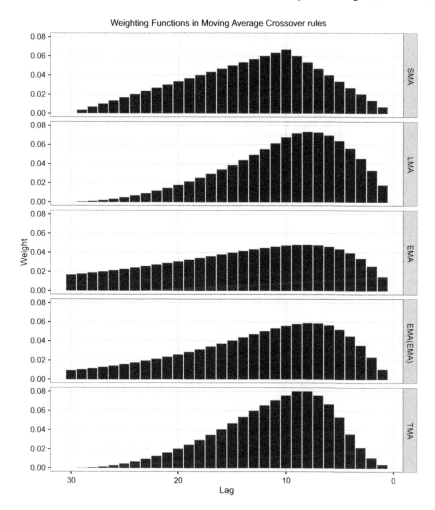

Fig. 5.3 The shapes of the price change weighting functions in five Moving Average Crossover rules: Simple Moving Average (SMA) Crossover rule, Linear (LMA) Moving Average Crossover rule, Exponential Moving Average (EMA) Crossover rule, Double Exponential Moving Average (EMA(EMA)) Crossover rule, and Triangular (TMA) Moving Average Crossover rule. In all rules, the sizes of the shorter and longer averaging windows equal $s = 10$ and $l = 30$ respectively

For the sake of illustration, Fig. 5.3 plots the shapes of the price change weighting functions in five Moving Average Crossover rules: Simple Moving Average (SMA) Crossover rule, Linear (LMA) Moving Average Crossover rule, Exponential Moving Average (EMA) Crossover rule, Double Exponential Moving Average (EMA(EMA)) Crossover rule, and Triangular (TMA) Moving Average Crossover rule. In all rules, the sizes of the shorter and longer averaging windows equal $s = 10$ and $l = 30$ respectively.

Recall from Sect. 4.5 that the Moving Average Crossover rule generates a much lesser number of false trading signals than the Price Minus Moving Average rule (at least, when daily data are used). In other words, the Moving Average Crossover rule reduces whipsaw trades. The foregoing analytic exposition revealed that, as compared to the price change weighting function of the Price Minus Moving Average rule, the price change weighting function of the Moving Average Crossover rule assigns lesser weights to the most recent price changes. This analytical result is supported by a visual comparison of the shapes of the price change weighting functions of some Moving Average Crossover rules (visualized in Fig. 5.3) and the shapes of the price change weighting functions of the corresponding Price Minus Moving Average rules (shown in Fig. 5.1). Consequently, the reduction in the number of false trading signals is achieved by reducing the weights of the most recent price changes. However, the reduction of weights of the most recent price changes has a side effect. Specifically, as compared to the Price Minus Moving Average rule, the Moving Average Crossover rule reacts with a longer delay to the changes in the price trend.

Traditionally, in the MAC(s, l) rule the size of the shorter averaging window is substantially smaller than the size of the longer averaging window, $s \ll l$. In this case the price change weighting function has a hump-shaped form where the top is located closer to the right end of the shape. However, the MAC(s, l) rule is very flexible and able to generate many different shapes of the price change weighting function. For the sake of illustration, Fig. 5.4 provides examples of possible shapes of the price change weighting functions generated by the Simple Moving Average Crossover rule. Specifically, when $s = 1$, the MAC(1, l) rule is equivalent to the P-MA(l) rule that assigns decreasing weights to more distant price changes. When $1 < s < l - 1$, the top of the hump-shaped form is located at lag s. If $s = l/2$, then the top of the hump-shaped form is located exactly in the middle of the averaging window. It is interesting to observe that, when $s = l - 1$, the price change weighting function assigns greater weights to more distant price changes. That is, the MAC(s, l) rule is able to produce both decreasing, humped, and increasing shapes of the price-change weighting function.

The illustrations of the shapes of the price change weighting functions in the Moving Average Crossover rule, provided in Figs. 5.3 and 5.4, are based on using moving averages with non-negative weights. However, there are moving averages, considered in Sect. 3.3, which assign negative weights to more distant prices in the averaging window. When moving averages have negative weights, the shape of the price change weighting function in the Moving Average Crossover rule becomes more elaborate. For the sake of illustration, Fig. 5.5

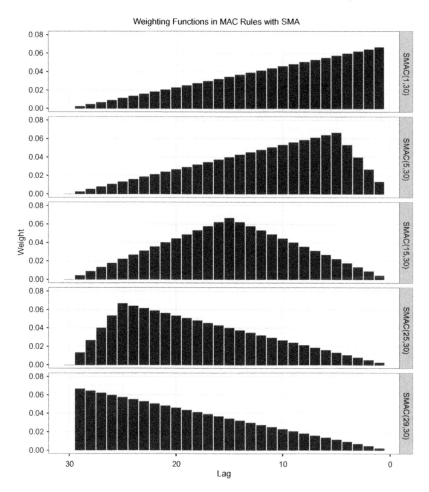

Fig. 5.4 The shapes of the price change weighting functions in five Simple Moving Average Crossover (SMAC) rules. In all rules, the size of the longer averaging window equals $l = 30$, whereas the size of the shorter averaging window takes values in $s \in [1, 5, 15, 25, 29]$.

plots the shapes of the price change weighting functions for the Moving Average Crossover rules based on the Double Exponential Moving Average (DEMA) and the Triple Exponential Moving Average (TEMA) proposed by Patrick Mulloy (see Mulloy 1994a, and Mulloy 1994b), the Hull Moving Average (HMA) proposed by Alan Hull (see Hull 2005), and the Zero Lag Exponential Moving Average (ZLEMA) proposed by Ehlers and Way (see Ehlers and Way 2010). Observe that all the price change weighting functions first increase, attain a maximum, then decrease below zero, attain a minimum, and finally increase toward zero. The pattern of the alternation of weights in these functions sug-

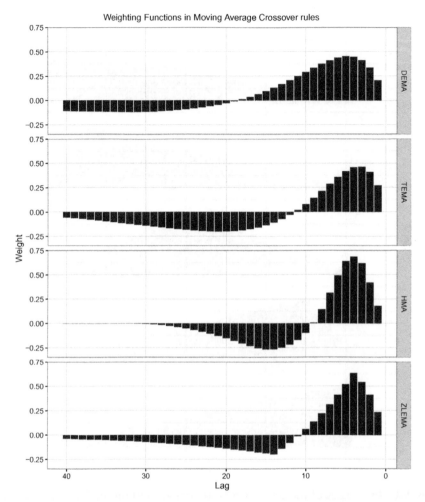

Fig. 5.5 The shapes of the price change weighting functions for the Moving Average Crossover rule based on the Double Exponential Moving Average (DEMA) and the Triple Exponential Moving Average (TEMA) proposed by Patrick Mulloy, the Hull Moving Average (HMA) proposed by Alan Hull, and the Zero Lag Exponential Moving Average (ZLEMA) proposed by Ehlers and Way. In all rules, the sizes of the shorter and longer averaging windows equal $s = 10$ and $l = 30$ respectively

gests that these rules are supposed to react to the changes in the price trend. For example, a strong Buy signal is generated when the prices first trend downward (the price changes are negative), then upward (the price changes are positive). Similarly, a strong Sell signal is generated when the prices first trend upward, then downward. Alternatively, these rules might work well when the prices are mean-reverting.

5.6   Moving Average Convergence/Divergence Rule

The computation of the technical trading indicator of the original MACD rule by Gerald Appel is based on using three Exponential Moving Averages:

$$MAC_t(s, l) = EMA_t(s) - EMA_t(l),$$

$$\text{Indicator}_t^{MACD(s,l,n)} = MAC_t(s, l) - EMA_t(n, MAC(s, l)).$$

For this rule, the computation of the trading indicator, in terms of price changes, is given by (see the subsequent appendix for the details of the derivation)

$$\text{Indicator}_t^{MACD(s,l,n)} = \sum_{j=1}^{\infty} \left(\left(\lambda_l^j - \lambda_s^j \right) - (1 - \lambda) \left[\frac{\lambda_l^j - \lambda^j}{1 - \frac{\lambda}{\lambda_l}} - \frac{\lambda_s^j - \lambda^j}{1 - \frac{\lambda}{\lambda_s}} \right] \right) \Delta P_{t-j},$$

$$(5.31)$$

where

$$\lambda_l = \frac{l - 1}{l + 1}, \quad \lambda_s = \frac{s - 1}{s + 1}, \quad \lambda = \frac{n - 1}{n + 1}.$$

Obviously, the computation of the trading indicator can also be interpreted as calculating the weighted average of price changes

$$\text{Indicator}_t^{MACD(s,l,n)} = \sum_{j=1}^{\infty} \pi_j \Delta P_{t-j}, \qquad (5.32)$$

where π_j is the weight of price change ΔP_{t-j} in the computation of the weighted average. However, in the case of the MACD rule, the weights π_j cannot be normalized because the sum of the weights equals zero (see the subsequent appendix for a proof).

Figure 5.6 illustrates the shapes of the price change weighting functions in three Moving Average Convergence/Divergence rules: the original MACD rule of Gerald Appel based on using Exponential Moving Averages (EMA), and two MACD rules of Patrick Mulloy based on using Double Exponential Moving Averages (DEMA) and Triple Exponential Moving Averages (TEMA). In all rules, the sizes of the averaging windows equal $s = 12$, $l = 26$, and $n = 9$ respectively.

The shape of the price change weighting function of the original MACD rule resembles the shape of the price change weighting function of the MAC

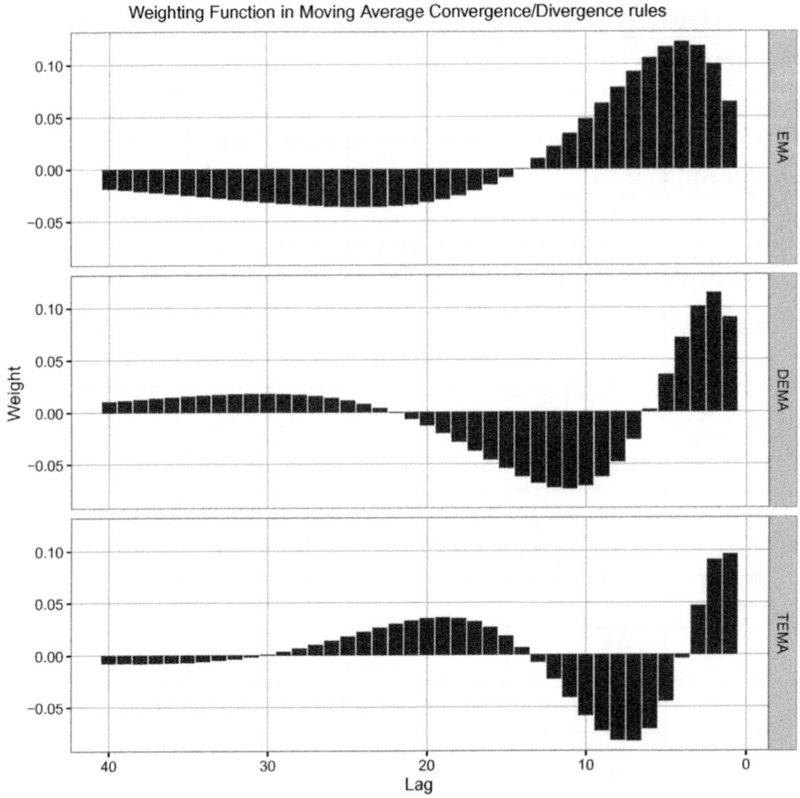

Weighting Function in Moving Average Convergence/Divergence rules

Fig. 5.6 The shape of the price change weighting functions in three Moving Average Convergence/Divergence rules: the original MACD rule of Gerald Appel based on using Exponential Moving Averages (EMA), and two MACD rules of Patrick Mulloy based on using Double Exponential Moving Averages (DEMA) and Triple Exponential Moving Averages (TEMA). In all rules, the sizes of the averaging windows equal $s = 12$, $l = 26$, and $n = 9$ respectively

rule where either DEMA or TEMA are used (see Fig. 5.5). The pattern of the alternation of weights in the original MACD rule confirms our observation made in Sect. 4.7. Specifically, the original MACD rule is designed to react to the changes in the price trend. The pattern of the alternation of weights in the two MACD rules of Patrick Mulloy resembles a damped harmonic oscillator (for example, a sine wave). This observation suggests that using either DEMA or TEMA in the MACD rule is sensible when prices are mean reverting with more or less stable period of mean-reversion.

5.7   Review of Anatomy of Trading Rules

This chapter demonstrates that the computation of a technical trading indicator for every moving average trading rule can alternatively be given by the following simple formula

$$\text{Indicator}_t^{TR(n)} = \sum_{i=1}^{n-1} \pi_i \Delta P_{t-i}. \tag{5.33}$$

In words, all technical trading indicators considered in this book are computed in the same general manner. In particular, any trading indicator is computed as a weighted average of price changes over the averaging window. As a result, any combination of a specific trading rule with a specific moving average of prices can be uniquely characterized by a peculiar weighting function of price changes. Therefore any differences between trading rules can be attributed solely to the differences between their price change weighting functions. As a natural consequence to this result, two seemingly different trading rules can be equivalent when their price change weighting functions are alike.

In spite of the fact that there is a great number of potential combinations of a specific trading rule with a specific moving average of prices, there are only four basic types (or shapes) of price change weighting functions:

1. Functions that assign equal weights to all price changes;
2. Functions that overweight (underweight) the most recent (distant) price changes;
3. Hump-shaped functions that underweight both the most recent and the most distant price changes;
4. Functions that have a damped waveform. Whereas in the previous types of weighting functions all price changes have non-negative weights, in this type the weights of price changes periodically change sign from positive to negative or vice versa.

The two trading rules that have equal weighting of price changes are the MOM rule (see Fig. 5.1) and the ΔSMA rule (see Fig. 5.2). The $\Delta SMA(n-1)$ rule is equivalent to the $MOM(n)$ rule.

The trading rules that overweight the most recent price changes include all P-MA rules based on moving averages with non-negative weights (see Fig. 5.1), as well as all ΔMA rules based on moving averages that overweight the most recent prices (see Fig. 5.2). The main examples of moving averages that over-weight the most recent prices are the LMA and the EMA. Both the P-SMA rule and the ΔLMA rule have a linear weighting function for price changes (see

Figs. 5.1 and 5.2). The $\Delta LMA(n-1)$ rule is equivalent to the $P - SMA(n)$ rule.

In a linear weighting function, the weights decrease linearly as the lag of a price change increases. Besides linear weighting, this type of a weighting function (that overweights the most recent price changes) can be a convex decreasing function, a concave decreasing function, or a decreasing function with several inflection points. We find that both the P-EMA rule and the ΔEMA rule have the same exponentially decreasing weighting function for price changes (again, see Figs. 5.1 and 5.2); hence, these two rules are equivalent. Another example of a trading rule with a convex decreasing weighting function for price changes is the P-LMA rule (see Fig. 5.1). The visual comparison of the price change weighting functions of the P-SMA and P-EMA rules (see Fig. 5.1) suggests that these two weighting functions look essentially similar; therefore we may expect that the performance of the P-SMA rule does not differ much from that of both the P-EMA and ΔEMA rules.

The hump-shaped weighting function for price changes can be created by using the MAC rule where both shorter and longer moving averages have only non-negative weights (see Fig. 5.3). The examples of such moving averages are all ordinary moving averages and moving averages of moving averages (where only ordinary moving averages are used). Alternatively, the hump-shaped weighting function for price changes can be created by using ΔMA rule based on a hump-shaped moving average (for example, ΔEMA(EMA) rule, see Fig. 5.2). Yet another way of creating a hump-shaped weighting function is to smooth the trading indicator, that employs a decreasing weighting function for price changes, using a shorter moving average. Since a decreasing weighting function can be created by either the P-MA or ΔMA rule, the two additional ways are

$$MA_s(P - MA_n) = MA_s - MA_s(MA_n),$$

and

$$MA_s(\Delta MA_n) = MA_s(MA_n) - MA_s(Lag_1(MA_n)).$$

The computation of the trading indicator of the $MA_s(P - MA_n)$ rule closely resembles the computation of the trading indicator of the $MAC(s, l)$ rule. Figure 5.7 demonstrates the shape of the price change weighting functions in three $MA_s(P - MA_n)$ rules that are based on SMA, LMA, and EMA. The shapes of these price change weighting functions closely resemble those of the price change weighting functions in the corresponding MAC rules (see Fig. 5.3).

The computation of the trading indicator of the $MA_s(\Delta MA_n)$ rule differs from the computation of the trading indicator of the $MAC(s, l)$ rule.

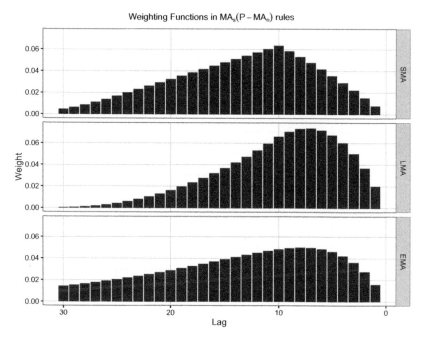

Fig. 5.7 The shape of the price change weighting functions in three $MA_s(P - MA_n)$ rules. In all rules, the sizes of the shorter and longer averaging windows equal $s = 10$ and $n = 26$ respectively

However, Fig. 5.8 shows that, when either LMA or EMA is used, the shapes of the price change weighting functions in the $MA_s(\Delta MA_n)$ rule closely resemble those of the price change weighting functions in the corresponding MAC rules (see Fig. 5.3). Only when SMA is used, the price change weighting function, even though it has a hump-shaped form, differs from the hump-shaped price-change weighting function of the MAC rule based on SMA (see Fig. 5.3).

The final type of a price change weighting function has a damped waveform. The main example of a trading rule that has this type of a price change weighting function is the MACD rule (see Fig. 5.6). However, the damped waveform of a price change weighting function can also be created by using the MAC rule based on moving averages that change sign (see Fig. 5.5). In particular, these moving averages assign positive weights to most recent prices, but negative weights to most distant prices.

The trading rules that have one of the first three types of the shape of the price change weighting function (equal, decreasing, or hump-shaped) are designed to identify the direction of the trend and generate a Buy (Sell) signal when prices trend upward (downward). These rules generate correct Buy and Sell trading signals when prices trend steadily upward or downward. However,

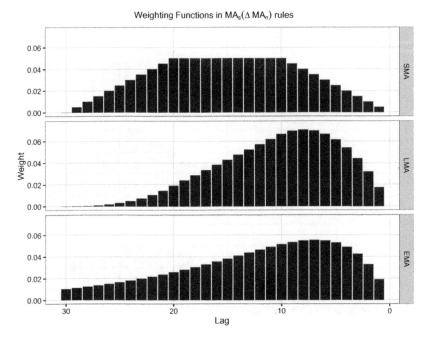

Fig. 5.8 The shape of the price change weighting functions in three $MA_s(\Delta MA_n)$ rules. In all rules, the sizes of the shorter and longer averaging windows equal $s = 10$ and $n = 30$ respectively

when prices go sideways, or the price trend often changes its direction, these rules do not work. In contrast, the trading rules that have the damped waveform shape of price change weighting function are designed to react to the changes in the trend direction. That is, these rules might be profitable when either the price trend often changes its direction or prices are mean-reverting. However, when prices trend steadily, these rules lose their advantage.

5.8   Chapter Summary

In this chapter we presented the methodology to study the computation of trading indicators in many market timing rules based on moving averages of prices and analyzed the commonalities and differences between the rules. Our analysis revealed that the computation of every technical trading indicator considered in this book can equivalently be interpreted as the computation of the weighted average of price changes over the averaging window. Despite a great variety of trading indicators that are computed seemingly differently at the first sight, we found that the only real difference between the diverse trading indicators lies in the weighting function used to compute the moving average of price changes. The most popular trading indicators employ either equal-weighting of price changes, overweighting the most recent price changes, a hump-shaped weighting function which underweights both the most recent and most distant price changes, or a weighting function that has a damped waveform where the weights of price changes periodically alter sign.

Our methodology of analyzing the computation of trading indicators for the timing rules based on moving averages offers a broad and clear perspective on the relationship between different rules. Whereas moving averages of prices are indispensable in visualizing how the trading signals are generated, because there is a great variety of trading rules, it is virtually impossible to see the commonalities and differences between various trading rules. In addition, if more than two moving averages are used to generate a trading signal, in this case it is also cumbersome to understand how a trading signal is generated. In contrast, our methodology of presenting the computation of the trading indicator in terms of a single moving average of price changes, rather than one or more moving averages of prices, uncovers the anatomy of trading rules and provides very useful insights about popular trend rules. In addition, our analysis offers a new and very insightful re-interpretation of the existing market timing rules.

The list of the useful insights about the popular trend rules, uncovered by our analysis, includes, but is not limited to, the following:

- Each trading rule based on one or multiple moving average of prices can be uniquely characterized by a single moving average of price changes.
- There are only four basic shapes of the weighting function for price changes.
- The same type of shape of the price change weighting function can be created using several alternative trading rules.
- There are trading rules with exactly the same shape of the price change weighting function; hence these rules are equivalent. The list of equivalent rules includes: the MOM and ΔSMA rules, the P-SMA and ΔLMA rules, and the P-EMA and ΔEMA rules.
- Virtually every trading rule can also be presented as a weighted average of the Momentum rules computed using different averaging periods. Thus, the Momentum rule might be considered as an elementary trading rule on the basis of which one can construct more elaborate rules.
- The trading rules that have either equal, decreasing, or hump-shaped form of the price change weighting function represent the "authentic"

trend rules. These rules are designed to generate correct signals when prices trend steadily upward or downward.

- The trading rules that have a damped waveform shape of the price change weighting function are designed to react to the changes in the trend direction. These rules generate correct signals when trend either accelerates or decelerates. Such rules might be profitable when either the price trend often changes its direction or prices are mean-reverting.

Table 5.1 summarizes the four main shapes of the price change weighting function and indicates which combinations of a specific trading rule with a specific type of moving average create which shape.

Table 5.1 Four main shapes of the price change weighting function in a trading rule based on moving averages of prices

Shape of weighting function	Trading rule	Moving average type
Equal weighting	MOM	
	ΔMA	SMA
Decreasing	P-MA	SMA, LMA, EMA, TMA, EMA(EMA)
	ΔMA	LMA, EMA
Hump-shaped	ΔMA	TMA, EMA(EMA)
	MAC	SMA, LMA, EMA, TMA, EMA(EMA)
Damped waveform	MAC	DEMA, TEMA, HMA, ZLEMA
	MACD	SMA, LMA, EMA, DEMA, TEMA

Notes This table summarizes the four main shapes of the price change weighting function and indicates which combinations of a specific trading rule with a specific type of moving average create which shape. For example, a decreasing price change weighting function (that overweights the most recent price changes) can be created by the Price Minus Moving Average (P-MA) rule where one of the following moving averages is used: Simple Moving Average (SMA), Linear Moving Average (LMA), Exponential Moving Average (EMA), Triangular Moving Average (TMA), and Exponential Moving Average of Exponential Moving Average (EMA(EMA)). As an another example, a price change weighting function that has a damped waveform can be created using the Moving Average Crossover (MAC) rule based on the following moving averages: Double Exponential Moving Average (DEMA), Triple Exponential Moving Average (TEMA), Hull Moving Average (HMA), and Zero Lag Exponential Moving Average (ZLEMA)

Appendix 5.A: Derivation of Formulas for Weighting Functions

5.A.1 Price Change Weighting Functions in the MAC rule

The general formula for the computation of the value of the technical trading indicator for the $MAC(s, l)$ rule

$$\text{Indicator}_t^{MAC(s,l)} = \frac{\sum_{j=1}^{l-1}\left(\sum_{i=j}^{l-1} w_i^l\right)\Delta P_{t-j}}{\sum_{i=0}^{l-1} w_i^l} - \frac{\sum_{j=1}^{s-1}\left(\sum_{i=j}^{s-1} w_i^s\right)\Delta P_{t-j}}{\sum_{i=0}^{s-1} w_i^s}.$$

(5.34)

If the Simple Moving Average is used (where $w_i = 1$ for all i) in both moving averages, then

$$\begin{aligned}
\text{Indicator}_t^{SMAC(s,l)} &= \frac{\sum_{j=1}^{l-1}\left(\sum_{i=j}^{l-1} 1\right)\Delta P_{t-j}}{\sum_{i=0}^{l-1} 1} - \frac{\sum_{j=1}^{s-1}\left(\sum_{i=j}^{s-1} 1\right)\Delta P_{t-j}}{\sum_{i=0}^{s-1} 1} \\
&= \frac{\sum_{j=1}^{l-1}(l-j)\Delta P_{t-j}}{l} - \frac{\sum_{j=1}^{s-1}(s-j)\Delta P_{t-j}}{s} \\
&= \sum_{j=1}^{s-1}\left(\frac{(l-j)}{l} - \frac{(s-j)}{s}\right)\Delta P_{t-j} + \sum_{j=s}^{l-1}\frac{(l-j)}{l}\Delta P_{t-j} \\
&= \sum_{j=1}^{s-1}\frac{(l-s)j}{l\times s}\Delta P_{t-j} + \sum_{j=s}^{l-1}\frac{(l-j)}{l}\Delta P_{t-j}.
\end{aligned}$$

(5.35)

Observe that the price change weighting function consists of two parts. From lag 1 to lag $s-1$, the price change weighting function is given by $\frac{(l-s)j}{l\times s}$. This price change weighting function increases when j increases because $l - s > 0$. From lag s till lag $l - 1$ the price change weighting function is given by $\frac{(l-j)}{l}$. This price change weighting function decreases when j increases. It is easy to check that the maximum weight is assigned to lag s. That is, the price change ΔP_{t-s} has the largest weight in the computation of the weighted average of price changes.

Now consider the computation of the technical trading indicator for the MAC rule where the Exponential Moving Average is used in both moving averages. Denote by λ_l and λ_s the decay factors in the longer and shorter moving averages respectively. Recall that $\lambda_l = \frac{l-1}{l+1}$ whereas $\lambda_s = \frac{s-1}{s+1}$. In this case the straightforward computations yield

$$
\begin{aligned}
\text{Indicator}_t^{\text{EMAC}(s,l)} &= \frac{\sum_{j=1}^{\infty}\left(\sum_{i=j}^{\infty}\lambda_l^i\right)\Delta P_{t-j}}{\sum_{i=0}^{\infty}\lambda_l^i} - \frac{\sum_{j=1}^{\infty}\left(\sum_{i=j}^{\infty}\lambda_s^i\right)\Delta P_{t-j}}{\sum_{i=0}^{\infty}\lambda_s^i} \\
&= \frac{\sum_{j=1}^{\infty}(1-\lambda_l)^{-1}\lambda_l^j\Delta P_{t-j}}{(1-\lambda_l)^{-1}} - \frac{\sum_{j=1}^{\infty}(1-\lambda_s)^{-1}\lambda_s^j\Delta P_{t-j}}{(1-\lambda_s)^{-1}} \\
&= \sum_{j=1}^{\infty}\left(\lambda_l^j - \lambda_s^j\right)\Delta P_{t-j}.
\end{aligned} \tag{5.36}
$$

As the result, in this case the price change weighting function is given by

$$
f(j) = \lambda_l^j - \lambda_s^j, \quad j \geq 1.
$$

This function is non-negative since $\lambda_l > \lambda_s$ (because $l > s$). As j increases, the function first increases, then decreases. To find the lag number at which the function attains ist maximum, we use the first-order condition for maximum

$$
f'(j) = \lambda_l^j \log(\lambda_l) - \lambda_s^j \log(\lambda_s) = 0.
$$

Solving this equation with respect to j yields

$$
j = \frac{\log\left(\frac{\log(\lambda_s)}{\log(\lambda_l)}\right)}{\log\left(\frac{\lambda_l}{\lambda_s}\right)}.
$$

5.A.2 Price Change Weighting Functions in the MACD rule

The computation of the technical trading indicator for the MACD rule is given by

$$
\text{Indicator}_t^{\text{MACD}(s,l,n)} = MAC_t(s,l) - EMA_t(n, MAC(s,l)),
$$

where $MAC_t(s,l)$ is the technical trading indicator for the MAC rule

$$
MAC_t(s,l) = EMA_t(s) - EMA_t(l),
$$

and $EMA_t(n, MAC(s, l))$ is the exponential moving average of the MAC trading indicator.

We know that, when Exponential Moving Averages are used, the computation of the technical trading indicator for the MAC rule can be written alternatively as

$$MAC_t(s, l) = \sum_{j=1}^{\infty} \left(\lambda_l^j - \lambda_s^j \right) \Delta P_{t-j}, \qquad (5.37)$$

where λ_l and λ_s denote the decay factors in the longer and shorter moving averages respectively ($\lambda_l = \frac{l-1}{l+1}$ whereas $\lambda_s = \frac{s-1}{s+1}$). The exponential moving average of the MAC trading indicator is computed as

$$EMA_t(n, MAC(s, l)) = (1 - \lambda) \sum_{i=0}^{\infty} \lambda^i MAC_{t-i}(s, l),$$

where $\lambda = \frac{n-1}{n+1}$ is the decay factor in the $EMA(n)$ and $MAC_{t-i}(s, l)$ is the lagged value of the MAC indicator given by

$$MAC_{t-i}(s, l) = \sum_{j=i+1}^{\infty} \left(\lambda_l^{j-i} - \lambda_s^{j-i} \right) \Delta P_{t-j}.$$

Therefore the computation of $EMA_t(n, MAC(s, l))$ can be written as

$$EMA_t(n, MAC(s, l)) = (1 - \lambda) \sum_{i=0}^{\infty} \lambda^i \left(\sum_{j=i+1}^{\infty} \left(\lambda_l^{j-i} - \lambda_s^{j-i} \right) \Delta P_{t-j} \right). \tag{5.38}$$

We proceed by rewriting expression (5.38) as

$$EMA_t(n, MAC(s, l)) = (1 - \lambda) \sum_{i=0}^{\infty} \left(\sum_{j=i+1}^{\infty} \left(\lambda_l^j \left(\frac{\lambda}{\lambda_l} \right)^i - \lambda_s^j \left(\frac{\lambda}{\lambda_s} \right)^i \right) \Delta P_{t-j} \right).$$

Interchanging the order of summation in the double sum above yields

$$
EMA_t(n, MAC(s, l)) = (1 - \lambda) \sum_{j=1}^{\infty} \left(\sum_{i=0}^{j-1} \left(\lambda_l^j \left(\frac{\lambda}{\lambda_l} \right)^i - \lambda_s^j \left(\frac{\lambda}{\lambda_s} \right)^i \right) \Delta P_{t-j} \right)
$$

$$
= (1 - \lambda) \sum_{j=1}^{\infty} \left[\lambda_l^j \left(\sum_{i=0}^{j-1} \left(\frac{\lambda}{\lambda_l} \right)^i \right) - \lambda_s^j \left(\sum_{i=0}^{j-1} \left(\frac{\lambda}{\lambda_s} \right)^i \right) \right] \Delta P_{t-j}.
$$

$$(5.39)$$

The closed-form expressions for the sums of the two geometric sequences in the formula above are given by

$$
\sum_{i=0}^{j-1} \left(\frac{\lambda}{\lambda_l} \right)^i = \frac{1 - \left(\frac{\lambda}{\lambda_l} \right)^j}{1 - \frac{\lambda}{\lambda_l}}, \quad \sum_{i=0}^{j-1} \left(\frac{\lambda}{\lambda_s} \right)^i = \frac{1 - \left(\frac{\lambda}{\lambda_s} \right)^j}{1 - \frac{\lambda}{\lambda_s}}.
$$

Therefore the resulting expression for $EMA_t(n, MAC(s, l))$ is as follows

$$
EMA_t(n, MAC(s, l)) = (1 - \lambda) \sum_{j=1}^{\infty} \left[\frac{\lambda_l^j - \lambda^j}{1 - \frac{\lambda}{\lambda_l}} - \frac{\lambda_s^j - \lambda^j}{1 - \frac{\lambda}{\lambda_s}} \right] \Delta P_{t-j}.
$$

$$(5.40)$$

Combining expressions (5.37) and (5.40) yields the final expression for the MACD rule

$$
\text{Indicator}_t^{MACD(s,l,n)} = \sum_{j=1}^{\infty} \left(\left(\lambda_l^j - \lambda_s^j \right) - (1 - \lambda) \left[\frac{\lambda_l^j - \lambda^j}{1 - \frac{\lambda}{\lambda_l}} - \frac{\lambda_s^j - \lambda^j}{1 - \frac{\lambda}{\lambda_s}} \right] \right) \Delta P_{t-j}.
$$

Again, we see that the computation of the technical indicator for a trading rule based on moving averages can be written as the weighted average of price changes

$$
\text{Indicator}_t^{MACD(s,l,n)} = \sum_{j=1}^{\infty} \pi_j \Delta P_{t-j}, \tag{5.41}
$$

where π_j is the weight of price change ΔP_{t-j} in the computation of the weighted average. However, the weights π_j cannot be normalized since the sum of the weights equals zero

$$
\begin{aligned}
\sum_{j=1}^{\infty} \pi_j &= \sum_{j=1}^{\infty} \left(\left(\lambda_l^j - \lambda_s^j \right) - (1-\lambda) \left[\frac{\lambda_l^j - \lambda^j}{1 - \frac{\lambda}{\lambda_l}} - \frac{\lambda_s^j - \lambda^j}{1 - \frac{\lambda}{\lambda_s}} \right] \right) \\
&= \frac{\lambda_l}{1-\lambda_l} - \frac{\lambda_s}{1-\lambda_s} - (1-\lambda) \left[\frac{\frac{\lambda_l}{1-\lambda_l} - \frac{\lambda}{1-\lambda}}{1 - \frac{\lambda}{\lambda_l}} - \frac{\frac{\lambda_s}{1-\lambda_s} - \frac{\lambda}{1-\lambda}}{1 - \frac{\lambda}{\lambda_s}} \right] \\
&= \frac{\lambda_l}{1-\lambda_l} - \frac{\lambda_s}{1-\lambda_s} - \left[\frac{\lambda_l}{\lambda_l - \lambda} \left(\frac{1-\lambda}{1-\lambda_l} \lambda_l - \lambda \right) - \frac{\lambda_s}{\lambda_s - \lambda} \left(\frac{1-\lambda}{1-\lambda_s} \lambda_s - \lambda \right) \right] \\
&= \frac{\lambda_l}{1-\lambda_l} - \frac{\lambda_s}{1-\lambda_s} - \left[\frac{\lambda_l}{1-\lambda_l} - \frac{\lambda_s}{1-\lambda_s} \right] = 0.
\end{aligned}
$$

References

Ehlers, J. F., & Way, R. (2010). Zero Lag (Well, Almost). *Technical Analysis of Stocks and Commodities, 28*(12), 30–35.

Hull, A. (2005). *How to reduce lag in a moving average.* http://www.alanhull.com/hull-moving-average, [Online; accessed 7-October-2016]

Mulloy, P. G. (1994a). Smoothing data with faster moving averages. *Technical Analysis of Stocks and Commodities, 12*(1), 11–19.

Mulloy, P. G. (1994b). Smoothing data with less lag. *Technical Analysis of Stocks and Commodities, 12*(2), 72–80.

Murphy, J. J. (1999). Technical analysis of the financial markets: A comprehensive guide to trading methods and applications. New York Institute of Finance.

Part III

Performance Testing Methodology

The Set of Tested Trading Rules and their Abbreviations

In the rest of the book we are going to test the profitability of moving average trading rules. A practical implementation of any trading rule, except the Momentum rule, requires choosing a particular type of moving average. Previously in this book we showed that there are many technical trading rules, as well as there are many popular types of moving averages. As a result, there exists a vast number of potential combinations of trading rules and moving averages of prices.

The detailed examination of the anatomy of moving average trading rules, presented in the preceding part of this book, suggests that any combination (of a trading rule and a moving average) can be uniquely characterized by a particular moving average of price changes. Luckily, despite a great number of potential combinations, there are only four basic shapes of the weighting function for price changes: equal, decreasing, humped form, and damped waveform. In order to generate these most typical shapes of the weighting function, we need, in principle, only three trading rules. Specifically, in the Momentum[1] rule ($MOM(n)$) all price changes have equal weights. The Moving Average Crossover rule ($MAC(s, l)$) is able to generate both the decreasing (when $s = 1$) and hump-shaped form (when $s > 1$) of the price-change weighting function. Finally, in the Moving Average Convergence/Divergence rule ($MACD(s, l, n)$), the shape of the price-change

---

[1]The Momentum rule is not a moving average trading rule in the conventional sense when one thinks in terms of moving averages of prices. However, when one thinks in terms of moving averages of price changes, the Momentum rule employs an equally-weighted average of price changes.

Table III.1 The set of trading rules, their abbreviations, and the shapes of their price-change weighting functions.

Trading rule	Moving average type			Shape of weighting function
	SMA	LMA	EMA	
MOM	-	-	-	Equally-weighted
P-MA	P-SMA	P-LMA	P-EMA	Decreasing
MAC	SMAC	LMAC	EMAC	Hump-shaped
MAE	SMAE	LMAE	EMAE	-
MACD	SMADC	LMACD	EMACD	Damped waveform

weighting function resembles a damped waveform. In addition to these three rules, we will also employ the Moving Average Envelope rule (MAE(n, p)). The reason for using this rule is that both the MAC and MAE rules are motivated by the same idea. Specifically, both of them are supposed to reduce the number of whipsaw trades in the Price Minus Moving Average rule (P-MA(n)).

In our study, we will use only ordinary moving averages: Simple Moving Average (SMA), Linear Moving Average (LMA), and Exponential Moving Average (EMA). This is because one does not need to employ exotic types of moving averages in order to generate a required shape of the price-change weighting function. Table III.1 lists the set of trading rules used in our study, their abbreviations, and the shape of the price-change weighting function in each rule. Note that the P-MA(n) rule is equivalent to the MAC(1, l) rule. Whereas the price-change weighting function in this rule has a decreasing shape, the MAC($s > 1, l > s$) rule usually generates a humped shape of the price-change weighting function. However, when the size of the shorter window s approaches the size of the longer window l, the price-change weighting function has an increasing shape. In contrast, when the size of the shorter window approaches 1, the price-change weighting function has a decreasing shape. That is, the MAC rule is able to generate three different shapes of the price-change weighting function. It is difficult to tell the shape of the price-change weighting function in the MAE($n, p > 0$) rule. However, as the boundaries of the envelope approach the moving average (when $p \rightarrow 0$), the price-change weighting function has a definite decreasing shape because the MAE(n, 0) rule is equivalent to the P-MA(n) rule. Finally, the shape of the price-change weighting function in the MACD rule resembles a dumped waveform.

6

Transaction Costs and Returns to a Trading Strategy

6.1 Transaction Costs in Capital Markets

In order to assess the real-life performance of a moving average trading strategy, we need to account for the fact that rebalancing an active portfolio incurs transaction costs. Transaction costs in capital markets consist of the following three main components: half-size of the quoted bid-ask spread, brokerage fees (commissions), and market impact costs. In addition there are various taxes applicable in some equity markets, delay costs, opportunity costs, etc. (see, for example, Freyre-Sanders et al. 2004). If investors sell securities they do not own (short sale), then they also incur short borrowing costs. All investors face the same bid-ask spreads and market-impact costs for a trade or short borrowing of any given size and security at any given moment. In contrast, the commissions (on purchase, sale, and short borrowing) are negotiated and depend on the annual volume of trading, as well as on the investor's other trading practices. In order to model realistic transaction costs, one usually distinguishes between two classes of investors (see, for example, Dermody and Prisman 1993): large (institutional) and small (individual).

Large investors are defined as those who frequently make large trades in blocks (of 10,000 shares) via the block trading desks or brokerage houses. Large investors usually face transaction costs schedule with no minimum fee specified. Large traders typically make ongoing agreements with the trading desks or brokerage houses to execute their trades for a flat institutional commission rate that applies to any volume of trade. Thus, commissions paid by large investors for trading a given stock are proportional to the number of shares traded. But marginal market impact costs for a given stock rise in the number of shares traded.

© The Author(s) 2017
V. Zakamulin, *Market Timing with Moving Averages*, New Developments
in Quantitative Trading and Investment, DOI 10.1007/978-3-319-60970-6_6

Small investors are defined as those who use retail brokerage firms and often trade in 100-share round lots. They can also trade odd lots. For small investors there is a minimum fee on any trade. They face retail commission rates for a given stock that decrease in the transaction size. Their total transaction costs, therefore, exhibit decreasing rates for any given trade up to some particular size. Individual investors pay substantially larger commissions than institutional investors. Specifically, whereas institutional investors usually pay very low commissions of about 0.1% (or even less) of the volume of trade, Hudson et al. (1996) report that individual investors pay commissions of about 0.5–1.5%.[1]

Market impact costs are closely related to liquidity: a relatively big order exerts pressure on price and, consequently, transaction costs increase with increasing order size. Market impact costs become a problem if an investor places an order to buy or sell a quantity of shares that is large relative to a market average daily share volume. Market impact costs are less significant with liquid stocks.[2] Liquidity refers to the ease with which a stock can be bought or sold without disturbing its price. Market impact costs are further considered to be the sum of two components: temporary and permanent price effects of trades.

Even such a brief review of the structure of transaction costs in capital markets reveals that it is not easy to model realistic transaction costs. The amount of transaction costs depends on the type of investor, liquidity of a financial asset, and the volume of trade. In addition, the bid-ask spread is higher during turbulent times and lower during calm times. That is, the bid-ask spread depends also on the volatility of a financial asset. To simplify the treatment of transaction costs, one usually assumes that transaction costs are proportional to the volume of trade. However, strictly speaking, this assumption is valid only for large investors who trade in liquid stocks. In this situation the quoted bid-ask spread is the main component of transaction costs and the market impact costs are negligible.

The formal treatment of the proportional transaction costs is as follows. We denote the bid price of the stock at time t by P_t^{bid} and the ask price by P_t^{ask} such that $P_t^{\text{bid}} < P_t^{\text{ask}}$. We suppose that P_t is the midpoint of the bid-ask prices, and we denote by τ the half-size of the ratio of the quoted bid-ask spread to the bid-ask price midpoint:

$$\tau = \frac{P_t^{\text{ask}} - P_t^{\text{bid}}}{2P_t}.$$

---

[1] However, commissions for individual investors have dropped a lot after 2000. Unfortunately, we do not have an updated reference on recent commissions.

[2] For example, large cap stocks are much more liquid than small cap stocks. As a result, not only the bid-ask spread for large cap stocks is less than that for small cap stocks, but also market impact costs for trading in large cap stocks are less than those for trading in small cap stocks.

Consequently, this allows us to interpret τ as proportional transaction costs such that

$$P_t^{\text{bid}} = (1 - \tau)P_t \text{ and } P_t^{\text{ask}} = (1 + \tau)P_t.$$

Observe that the commissions that are proportional to the volume of trade can easily be incorporated in τ.

For our study we need to find estimates of the average one-way transaction costs in various capital markets. The problem is that the financial literature reports different estimates of the average one-way transaction costs in stock markets. Specifically, on the one hand, Berkowitz et al. (1988), Chan and Lakonishok (1993), and Knez and Ready (1996) estimate the average one-way transaction costs for institutional investors to be 0.25%. On the other hand, Stoll and Whaley (1983), Bhardwaj and Brooks (1992), Lesmond et al. (1999), Balduzzi and Lynch (1999), and Bessembinder (2003) document that the average one-way transaction costs amount to 0.50%.[3]

The government bonds are more liquid securities as compared to stocks and, therefore, the average bid-ask spread in bond trading is smaller than that in stock trading. Chakravarty and Sarkar (2003) and Edwards et al. (2007) estimate the average one-way transaction costs in trading intermediate- and long-term bonds to be about 0.10%. Finally, the US Treasury Bills of maturities of 1–3 months are highly liquid securities with virtually zero bid-ask spread. Therefore one usually assumes that buying and selling Treasury Bills is costless.

6.2 Computing the Returns to a Trading Strategy

The process of generation of a trading signal in all moving average trading rules is considered in Sect. 4.1. In brief, denoting the time t value of a technical trading indicator by Indicator$_t$, a Buy signal is generated when the value of the technical trading indicator is positive. Otherwise, a Sell signal is generated. That is,

$$\text{Signal}_{t+1} = \begin{cases} \text{Buy} & \text{if Indicator}_t > 0, \\ \text{Sell} & \text{if Indicator}_t \leq 0. \end{cases}$$

---

[3]Again, these references are probably outdated because after 2000 the liquidity in the stock markets has improved. Unfortunately, we do not have updated estimates on the average bid-ask spread in the stock markets. Therefore in our tests we employ the lower estimate for the average one-way transaction costs of 0.25%.

Let $(R_1, R_2, \ldots, R_T)$ be the (total) returns on stocks, and let $(r_{f1}, r_{f2}, \ldots, r_{fT})$ be the risk-free rates of return over the same sample period. A Buy signal is always a signal to invest in the stocks (or stay invested in the stocks). When a Sell signal is generated, there are two alternative strategies. Most commonly, a Sell signal is a signal to sell the stocks and invest the proceeds in cash (or stay invested in cash). In this case, in the presence of transaction costs, the return to the market timing strategy over $t + 1$ is given by

$$
r_{t+1} = \begin{cases}
R_{t+1} & \text{if (Signal}_{t+1} = \text{Buy) and (Signal}_t = \text{Buy)}, \\
R_{t+1} - \tau & \text{if (Signal}_{t+1} = \text{Buy) and (Signal}_t = \text{Sell)}, \\
r_{ft+1} & \text{if (Signal}_{t+1} = \text{Sell) and (Signal}_t = \text{Sell)}, \\
r_{ft+1} - \tau & \text{if (Signal}_{t+1} = \text{Sell) and (Signal}_t = \text{Buy)},
\end{cases} \tag{6.1}
$$

where, recall, τ denotes the average one-way transaction costs in trading stocks and we assume that trading in the risk-free asset is costless. Note that if the signal was Buy during the previous period and the signal is Buy for the subsequent period, then the return to the moving average strategy (over the subsequent period) equals the return on stocks. If the signal was Sell during the previous period, money was kept in cash. When a Buy signal is generated, a trader must buy stocks and therefore the return to the moving average strategy equals the return on stocks less the amount of transaction costs.[4] Similarly, if the signal was Sell during the previous period and the signal is Sell for the subsequent period, the return to the moving average strategy equals the risk-free rate of return. If the signal was Buy during the previous period, money was invested in stocks. When a Sell signal is generated, a trader must sell stocks and therefore the risk-free rate of return for the subsequent period is reduced by the amount of transaction costs.

Short selling stocks means borrowing some number of shares of a stock with subsequent selling these shares in the market. At some later point in time the short-seller must buy back the same number of shares and return them to the lender. In the strategy where a trader shorts stocks when a Sell signal is generated, the amount of transaction costs doubles. This is because, when a Buy signal is generated after a Sell signal, a trader needs to buy some number of shares of the stock in order to return them to the lender, and additionally buy the same number of shares of the stock for personal investment. Similarly, when a Sell signal is generated after a Buy signal, a trader needs to sell all own shares of the stock and, right after selling own shares, sell short the same

---

[4]More exactly, since the transaction takes place at the close ask price $P_t^{\text{ask}} = (1 + \tau)P_t$, the return to the moving average strategy equals $\frac{R_{t+1} - \tau}{1 + \tau}$. However, since $1 + \tau \approx 1$, the expression $R_{t+1} - \tau$ closely approximates the real return.

number of shares of the stock. The proceeds from the sale and the short sale are invested in cash and, as a result, during the period when the stocks are sold short, the trader's return equals twice the return on the risk-free asset. Overall, in the presence of transaction costs, in this case the return to the moving average strategy over $t + 1$ is given by

$$
r_{t+1} = \begin{cases}
R_{t+1} & \text{if } (\text{Signal}_{t+1} = \text{Buy}) \text{ and } (\text{Signal}_t = \text{Buy}), \\
R_{t+1} - 2\tau & \text{if } (\text{Signal}_{t+1} = \text{Buy}) \text{ and } (\text{Signal}_t = \text{Sell}), \\
2r_{ft+1} - R_{t+1} & \text{if } (\text{Signal}_{t+1} = \text{Sell}) \text{ and } (\text{Signal}_t = \text{Sell}), \\
2r_{ft+1} - R_{t+1} - 2\tau & \text{if } (\text{Signal}_{t+1} = \text{Sell}) \text{ and } (\text{Signal}_t = \text{Buy}).
\end{cases}
\tag{6.2}
$$

6.3   Chapter Summary

Following a passive buy-and-hold strategy involves no trading. However, every active portfolio strategy requires continuous monitoring the market dynamics and sometimes frequent rebalancing the composition of the active portfolio. Even when the amount of transaction costs is relatively small, frequent trading may incur large transaction costs and seriously deteriorate the performance of the active strategy. Thus, transaction costs represent a very important market friction that must be seriously taken into account while assessing the real-life performance of a trading strategy. Unfortunately, the amount of transaction costs is difficult to estimate because it depends on many variables. Therefore, for the sake of simplicity, one usually assumes that transaction costs are linearly proportional to the volume of trade. However, even under this simplified assumption it is very difficult to estimate the average transaction costs. In stock markets, the estimate for the average one-way transaction costs varies from 0.25 to 0.50% (25 to 50 basis points). On the bright side, the simplified treatment of transaction costs allows one to easily incorporate the transaction costs in the returns to the simulated trading strategy.

References

Balduzzi, P., & Lynch, A. W. (1999). Transaction costs and predictability: Some utility cost calculations. *Journal of Financial Economics, 52*(1), 47–78.

Berkowitz, S. A., Logue, D. E., & Noser, E. A. (1988). The total costs of transactions on the NYSE. *Journal of Finance, 43*(1), 97–112.

Bessembinder, H. (2003). Issues in assessing trade execution costs. *Journal of Financial Markets, 6*(3), 233–257.

Bhardwaj, R. K., & Brooks, L. D. (1992). The January anomaly: Effects of low share price, transaction costs, and bid-ask bias. *Journal of Finance, 47*(2), 553–575.

Chakravarty, S., & Sarkar, A. (2003). Trading costs in three U.S. bond markets. *Journal of Fixed Income, 13*(1), 39–48.

Chan, L. K. C., & Lakonishok, J. (1993). Institutional trades and intraday stock price behavior. *Journal of Financial Economics, 33*(2), 173–199.

Dermody, J. C., & Prisman, E. Z. (1993). No arbitrage and valuation in markets with realistic transaction costs. *Journal of Financial and Quantitative Analysis, 28*(1).

Edwards, A. K., Harris, L. E., & Piwowar, M. S. (2007). Corporate bond market transaction costs and transparency. *Journal of Finance, 62*(3), 1421–1451.

Freyre-Sanders, A., Guobuzaite, R., & Byrne, K. (2004). A review of trading cost model: Reducing transaction costs. *Journal of Investing, 13*(3), 93–116.

Hudson, R., Dempsey, M., & Keasey, K. (1996). A note on the weak form efficiency of capital markets: The application of simple technical trading rules to UK stock prices-1935 to 1994. *Journal of Banking and Finance, 20*(6), 1121–1132.

Knez, P. J., & Ready, M. J. (1996). Estimating the profits from trading strategies. *Review of Financial Studies, 9*(4), 1121–1163.

Lesmond, D. A., Ogden, J. P., & Trzcinka, C. A. (1999). A new estimate of transaction costs. *Review of Financial Studies, 12*(5), 1113–1141.

Stoll, H. R., & Whaley, R. E. (1983). Transaction costs and the small firm effect. *Journal of Financial Economics, 12*(1), 57–79.

7

Performance Measurement and Outperformance Tests

7.1 Choice Under Uncertainty and Portfolio Performance Measures

Using the historical data for the returns to the buy-and-hold strategy (for example, the returns on a broad stock market index), $\{R_t\}$, and the risk-free rates of return, $\{r_{ft}\}$, the investor can easily simulate the returns to some particular moving average trading strategy $\{r_t\}$. The next problem is more difficult: by comparing the properties of the two return series, $\{R_t\}$ and $\{r_t\}$, the investor needs to decide which strategy performed better than the other. Unfortunately, there is no unique solution to this paramount problem because of the uncertainty involved. That is, following each strategy involves risk taking; each strategy can be considered as a distinct risky asset.

In the subsequent exposition, we briefly review how the choice of the best risky asset (or a portfolio) is done within the framework of modern finance theory. To generalize the exposition, we consider the investor's choice between two mutually exclusive risky portfolios A and B whose returns are denoted by r_A and r_B respectively. In addition to the risky assets, finance theory usually assumes the existence of a risk-free (or safe) asset. The interest rate on a short-term Treasury Bill commonly serves as a proxy for the risk-free rate of return denoted by r_f. The role of the risk-free asset is to control the risk of the investor's complete portfolio[1] through the fraction of wealth invested in the safe asset. It is usually assumed that the investor can either borrow or save at the risk-free rate and borrowing is not limited.

---

[1] In our exposition, we closely follow the exposition and terminology used in the introductory text on investments by Bodie et al. (2007).

© The Author(s) 2017
V. Zakamulin, *Market Timing with Moving Averages*, New Developments in Quantitative Trading and Investment, DOI 10.1007/978-3-319-60970-6_7

The investor's "capital allocation" consists of investing proportion a in the risky asset r_i ($i \in \{A, B\}$) and, consequently, $1 - a$ in the risk-free asset. The return on the investor's complete portfolio is given by

$$r_c^i = a\, r_i + (1 - a)r_f = a(r_i - r_f) + r_f. \qquad (7.1)$$

Notice that if $0 < a < 1$, the investor splits the wealth between the risky and the risk-free asset. If $a = 1$, the investor's wealth is placed in the risky asset only. Finally, if $a > 1$, the investor borrows money at the risk free rate and invests all own money and borrowed money in the risky asset.

If the investor chooses asset A, the investor's final wealth is given by

$$W_A = W_0(1 + r_c^A),$$

where W_0 denotes the investor's initial wealth. Similarly, if the investor chooses asset B, the investor's final wealth is given by

$$W_B = W_0(1 + r_c^B).$$

If the returns r_A and r_B were deterministic (that is, certain), then the choice of the best asset would be very simple. In particular, the best asset would be the asset which provides the highest rate of return.[2] The choice of the best asset becomes much more complicated when the returns are uncertain. As a result, portfolio performance evaluation is a lively research area within modern finance theory. Researchers have proposed a vast number of different portfolio performance measures (see Cogneau and Hübner 2009, for a good review of different performance measures). By a performance measure in finance one means a score attached to each risky portfolio. This score is usually used for the purpose of ranking of risky portfolios. That is, the higher the performance measure of a portfolio, the higher the rank of this portfolio. The goal of any investor who uses a particular performance measure is to select the portfolio for which this measure is the greatest. Most of the proposed performance measures are so-called "reward-to-risk" ratios. Below we review a few popular portfolio performance measures and point to their advantages and disadvantages.

---

[2]It should be noted, however, that the existence of two assets with deterministic but different returns is impossible because it creates profitable arbitrage opportunities.

7.1.1   Mean Excess Return

At first sight it seems rather straightforward to assume that, when returns are uncertain, the investor's natural goal might be to choose the asset which maximizes the expected future wealth. That is, the investor can compare $E[W_A]$ and $E[W_B]$, where $E[\cdot]$ denotes the expectation operator, and choose the asset which provides the highest future expected wealth. In this case one can use the mean excess return, $E[r_i - r_f]$ as a performance measure.

However, a closer look at this measure reveals a serious problem that consists of the following. Since we assume that the investor's goal is to maximize the future expected wealth, the investor has to solve the following optimal capital allocation problem

$$\max_a E[W_0(1 + a(r_i - r_f) + r_f)].$$

When $E[r_i - r_f] > 0$ and borrowing at the risk-free interest rate is not limited, there is no solution to this problem because the higher the value of a, the greater the investor's future expected wealth. If the investor behaves as though his objective function is to maximize the future expected wealth, such an investor would be willing to borrow an infinite amount at the risk-free rate and invest it in the risky asset. Thus, the mean excess return decision criterion produces a paradox. In particular, a seemingly sound criterion predicts a course of action that no actual investor would be willing to take.

The mean excess return of a risky asset, often termed as the "reward" measure, is an important measure that characterizes the properties of a risky asset. The other important characteristic of a risky asset is its measure of risk. The paradox presented above appears because we assume that in making financial decisions the investor ignores risk. When we assume that the goal of each investor is to choose a risky asset that provides the best tradeoff between the risk and reward, we arrive to a so-called "reward-to-risk" measure. Two of such measures are considered below in the subsequent sections.

To recap, the great disadvantage of the mean excess return performance measure is the ignorance of risk. However, because the notion of "risk" is an ambiguous concept, the ignorance of risk makes this measure independent of the investor's risk preferences. Besides, the mean return criterion, $E[r_i]$, can also be used in the absence of a risk-free asset. This is advantageous because all other rational reward-to-risk measures are constructed assuming the existence of a risk-free asset. When there is no risk-free asset, the arguments behind the construction of rational reward-to-risk measures break down. Last but not least, the rationale behind using the mean excess return measure is that in real markets the borrowing is limited and, when it comes to individual investors, often just impossible. When borrowing at the risk-free rate is limited or impossible, the paradox produced by the mean excess return decision criterion disappears.

7.1.2 Sharpe Ratio

Modern financial theory suggests that the choice of the best risky asset depends on the investor's risk preferences that are generally described by a utility function defined over investor's wealth. Unfortunately, the expected utility theory (originally presented by von Neumann and Morgenstern 1944) is silent about the shape of the investor's utility function. The standard assumptions in finance are that the utility function is increasing and concave in wealth. Still, there are plenty of mathematical functions that satisfy these assumptions.

Under certain additional simplified assumptions,[3] the investor's utility function can be approximated by the mean-variance utility

$$U(W) = E[W] - \frac{1}{2}A \times Var[W], \tag{7.2}$$

where $Var[W]$ is the variance of wealth and A is the investor's coefficient of risk aversion. It can be shown further that the mean-variance utility can equivalently be computed over returns (see Bodie et al. 2007)

$$U(r_c) = E[r_c] - \frac{1}{2}A\,\sigma_c^2,$$

where $E[r_c]$ and σ_c^2 denote the mean and variance of returns, respectively, of the investor's complete portfolio. In this form, the investor's utility function motivates using the variance (or standard deviation) of returns as a risk measure.

The mean and standard deviation of the investor's complete portfolio (see Eq. 7.1) are given by

$$E[r_c^i] = aE[r_i - r_f] + r_f, \quad \sigma_c^i = a\sigma_i.$$

---

[3]The use of the mean-variance utility function can be justified when either return distributions are normal or the investor is equipped with the quadratic utility function, see Tobin (1969) and Levy and Markowitz (1979).

The combination of these two equations yields the following relationship between the expected return and the risk of the complete portfolio:

$$E[r_c^i] = r_f + \frac{E[r_i - r_f]}{\sigma_i}\sigma_c^i. \tag{7.3}$$

Equation (7.3) says that there is a linear relation between the mean and standard deviation of returns of the investor's complete portfolio. In the standard deviation - mean return space, this strait line is called the Capital Allocation Line (CAL). It depicts all risk-return combinations available to investors who allocate wealth between the risk-free asset and risky asset i. The intercept and the slope of the straight line equal r_f and $\frac{E[r_i - r_f]}{\sigma_i}$ respectively.

William Sharpe (see Sharpe 1966, and Sharpe 1994) was the first to observe that, in the mean-variance framework where investors can borrow and lend at the risk-free rate, the choice of the best risky asset does not depend on the investor's attitude toward risk. Specifically, all investors regardless of their levels of risk aversion choose the same risky asset: the asset with the highest slope of the capital allocation line. Therefore the slope of the capital allocation line can be used to measure the performance of a risky asset (or portfolio). William Sharpe originally called this slope as "reward-to-variability" ratio. Later this ratio was termed the "Sharpe ratio":

$$SR_i = \frac{E[r_i - r_f]}{\sigma_i}.$$

For the sake of illustration, Fig. 7.1 indicates the locations of two risky assets, A and B, and the risk-free asset in the standard deviation - mean return space. Notice that, as compared to asset A, asset B provides a higher mean return with higher risk. Without the presence of the risk-free asset the choice the best risky asset depends on the investor's coefficient of risk aversion. More risk averse investors tend to prefer asset A to asset B, whereas more risk tolerant investor tend to prefer asset B to asset A. However, in the presence of the risk-free asset the choice of the best risky asset is unique. Since the slope of the capital allocation line through A is higher than that through B, all investors prefer asset A to asset B. To realize this, suppose that the investor wants to attain some arbitrary level of mean returns r^*. If the investor chooses asset A for capital allocation, the risk-return combination of the investor's complete portfolio is given by point "a" that belongs to the capital allocation line through asset A. In contrast, if the investor chooses asset B for capital allocation, the risk-return combination of the investor's complete portfolio is given by point "b" that belongs to the capital allocation line through asset B. Obviously, since

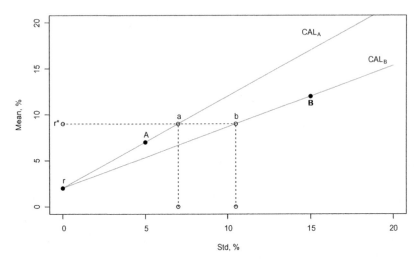

Fig. 7.1 The standard deviation - mean return space and the capital allocation lines (CALs) through the risk-free asset r and two risky assets A and B

both combinations, "a" and "b", have the same mean return but "a" is less risky than "b", any investor prefers "a" to "b". Consequently, any investor chooses asset A.

Even though the Sharpe ratio is a routinely used performance measure in the situations where the investor has to choose a single risky asset from a universe of many mutually exclusive risky assets, one has to keep in mind that the justification of the usage of this ratio is based on many assumptions that can be violated in reality:

- When return distributions are asymmetrical, the risk cannot be adequately measured by standard deviation that penalizes equally losses and gains;
- In reality, borrowing at the risk-free rate is either restricted or just impossible. In this case the investor cannot attain any arbitrary level of mean returns. For example, in the illustration on Fig. 7.1 the investor cannot attain r^* using asset A in the capital allocation. As a consequence, risk tolerant investors tend to prefer asset B even though it has a lower Sharpe ratio;
- The assumption about the existence of a risk-free asset is very crucial. Without the existence of a risk-free asset the choice of the best risky asset is not unique. Strictly speaking, there are no risk-free assets in reality. For

example, either the government that issues Treasury Bills may default, or the investor has a long and uncertain investment horizon.

Last but not least, keep in mind that the goal of any investor is to maximize the expected utility of future wealth. To attain this goal, investors need to solve not one, but two optimization problems at the same time: (1) to choose the optimal risky asset and (2) to choose the optimal capital allocation. If the risky asset is chosen optimally, but the capital allocation is not optimal, the investor fails to maximize the expected utility. Consequently, in some situations, by using an inferior risky asset but allocating capital optimally, the investor can achieve higher expected utility as compared to the case when the risky asset is chosen optimally but is used in far from optimal capital allocation.

7.1.3   Sortino Ratio

The Sharpe ratio is often criticized on the grounds that the standard deviation is not an adequate risk measure. In particular, the standard deviation penalizes similarly both the downside risk and upside return potential. Many researchers and practitioners argue that a proper risk measure must take into account only downside risk. This argument might be relevant in our context. Specifically, since a market timing strategy is supposed to provide downside protection and upside participation, the use of the Sharpe ratio for performance measurement of market timing strategies might be inappropriate.

The most known reward-to-risk performance measure that takes into account only the downside risk is the Sortino ratio (see Sortino and Price 1994). Originally, the Sortino ratio was presented as an ad-hoc performance measure. Subsequently, Pedersen and Satchell (2002) and Zakamulin (2014) presented a utility-based justification of the Sortino ratio. In particular, these authors showed that the Sortino ratio is a performance measure of investors that have a mean-downside variance utility function.[4] This utility function is similar to the mean-variance utility function where variance σ^2 is replaced by downside variance θ^2. The downside variance of risky asset r_i is computed as

$$\theta_i^2 = E\left[\min(r_i - r_f, 0)^2\right].$$

---

[4]It is worth noting that, whereas the mean-variance utility function can be justified on the grounds of expected utility theory, the mean-downside variance utility function can be justified on the grounds of behavioral finance theory, see Zakamulin (2014).

Note that the downside variance is defined as the expected square deviation below the risk-free rate of return. The resulting utility function is given by

$$U(r_c) = E[r_c] - \frac{1}{2} A \theta_c^2.$$

The mean and downside standard deviation of the investor's complete portfolio are given by

$$E[r_c^i] = a E[r_i - r_f] + r_f, \quad \theta_c^i = a\theta_i.$$

The combination of these two equations yields the following relationship between the expected return and the risk of the complete portfolio:

$$E[r_c^i] = r_f + \frac{E[r_i - r_f]}{\theta_i} \theta_c^i. \tag{7.4}$$

In the downside standard deviation - mean return space, this strait line can be again called the Capital Allocation Line (CAL) that depicts all risk-return combinations available to investors who allocate wealth between the risk-free asset and risky asset i. As in the case where the risk is measured by standard deviation, in the presence of the risk-free asset the choice of the best risky asset does not depend on the investor's risk preferences when the risk is measured by downside standard deviation. The best risky asset is the asset with the highest slope of the capital allocation line. This slope is best known as the "Sortino ratio":

$$SoR_i = \frac{E[r_i - r_f]}{\theta_i}.$$

It should be noted, however, that in the original definition of the Sortino ratio (made by Sortino and Price 1994) the downside variance is computed using an arbitrary return level k instead of the risk-free rate of return. That is, in the original definition the downside variance is computed as $E[\min(r_i - k, 0)^2]$. The problem is that when $k \neq r_f$, the capital allocation line is not a straight line in the risk-reward space. As a result, the choice of the best risky asset becomes dependent on the investor's risk preferences.

As a final remark, it is worth mentioning that the only potential advantage of the Sortino ratio over the Sharpe ratio is that the former employs a downside risk measure. The Sortino ratio retains all the other weaknesses of the Sharpe ratio. Specifically, the arguments that justify the use of the Sortino ratio break down when either the borrowing at the risk-free rate is restricted or the risk-free asset does not exist.

7.2   Statistical Tests for Outperformance

7.2.1   Estimating Performance Measures

Denote by $\{r_t\}$ the series of returns to a moving average trading strategy over some historical sample of size T. Over the same sample, the series of returns to the buy-and-hold strategy and the risk-free rates of returns are given by $\{R_t\}$ and $\{r_{ft}\}$ respectively. Note that all performance measures presented in this section are computed using the excess returns[5]

$$r_t^e = r_t - r_{ft}, \quad R_t^e = R_t - r_{ft}.$$

The mean excess return, the standard deviation of excess returns, and the downside standard deviation of the moving average trading strategy are estimated using the following formulas:

$$\bar{r}_{MA} = \bar{r}^e = \frac{1}{T}\sum_{t=1}^{T} r_t^e, \quad \hat{\sigma}_{MA} = \sqrt{\frac{1}{T-1}\sum_{t=1}^{T}(r_t^e - \bar{r}^e)^2}, \quad \hat{\theta}_{MA} = \sqrt{\frac{1}{T-1}\sum_{t=1}^{T}\min(r_t^e, 0)^2}.$$

Subsequently, the Sharpe and Sortino ratios of the moving average trading strategy are computed according to:

$$\widehat{SR}_{MA} = \frac{\bar{r}_{MA}}{\hat{\sigma}_{MA}}, \quad \widehat{SoR}_{MA} = \frac{\bar{r}_{MA}}{\hat{\theta}_{MA}}.$$

Similarly, the mean excess return, the Sharpe and Sortino ratios of the buy-and-hold strategy are estimated. These performance measures are denoted by $\bar{r}_{BH}$, $\widehat{SR}_{BH}$, and $\widehat{SoR}_{BH}$ respectively.

Observe that a "bar" is placed over the mean excess return to indicate that this is an estimator of the mean excess return, not the true value of the mean excess return (for example, $\bar{r}_{MA}$ is an estimator of r_{MA}). Similarly, a "hat" is placed over the standard deviation, downside standard deviation, the Sharpe and Sortino ratios to indicate that all these values are estimators, not the true values (for example, $\widehat{SR}_{MA}$ is an estimator of SR_{MA}).

---

[5]See Sharpe (1994) who advocates that the standard deviation in the Sharpe ratio should be computed using the excess returns.

7.2.2 Formulating the Outperformance Hypothesis

Denote by $\widehat{\mathcal{M}}_{MA}$ and $\widehat{\mathcal{M}}_{BH}$ the estimated performance measures of the moving average trading strategy and the corresponding buy-and-hold strategy. The first step in evaluating, whether the performance of the moving average strategy is higher than the performance of the buy-and-hold strategy, is to subtract the performance measure of the buy-and-hold strategy from the performance measure of the moving average strategy. That is, to compute the following difference that we call the "outperformance":

$$\widehat{\Delta} = \widehat{\mathcal{M}}_{MA} - \widehat{\mathcal{M}}_{BH}.$$

Suppose that $\widehat{\Delta} > 0$. Can we conclude on this information alone that the moving average strategy outperforms its passive counterpart? The answer to this question is, in fact, negative. This is because the time series $\{r_t^e\}$ and $\{R_t^e\}$ can be considered as series of observations of two random variables. As a result, the estimator $\widehat{\Delta}$ is also a random variable and the outperformance (the observation of $\widehat{\Delta} > 0$) can appear due to chance. To evaluate whether the moving average strategy produces "true" outperformance, we need to carry out a statistical test to see if the value of $\widehat{\Delta}$ is statistically significantly above zero. For this purpose we formulate the following null and alternative hypotheses about the true value of outperformance (denoted by Δ):

$$H_0 : \Delta \leq 0 \text{ versus } H_A : \Delta > 0. \tag{7.5}$$

In our context, a statistical hypothesis is a conjecture about the true value of Δ. Note that any hypothesis test involves formulating two hypothesis: one is called "null hypothesis" (denoted by H_0) and the other "alternative hypothesis" (denoted by H_A). Both of the two hypotheses are defined as mutually exclusive. A hypothesis test is a formal statistical procedure for deciding which of the two alternatives, H_0 or H_A, is more likely to be correct. The result of a hypothesis test leads to one of two decisions: either reject H_0 (in favor of H_A) or retain H_0. The decision "to reject or not to reject" H_0 depends on how likely H_0 to be true.

The idea behind testing our hypothesis is as follows. Denote by δ a numerical outcome of the random variable Δ. First, we learn the probability distribution of Δ under the null hypothesis. As the result, we know the probability that the random variable Δ takes on the particular value δ. If H_0 is true, then a random outcome $\delta \geq \widehat{\Delta}$ (under condition that $\widehat{\Delta} > 0$) would rarely happen. Consequently, the result of our hypothesis test is the probability of observing $\delta \geq \widehat{\Delta}$ under the null hypothesis. This probability is commonly called the "p-value". For example, suppose that $\widehat{\Delta} = 0.2$ and the p-value of the test equals 3%. This means that, assuming the null hypothesis were true, the probability of observing $\delta \geq 0.2$ equals 3%, which is highly unlikely. Therefore we can reject H_0 in favor of H_A. If, on the other hand, the p-value of the test equals 30%, it means that the probability of observing $\delta \geq 0.2$ equals 30% which is not "unusual enough". In this case we cannot reject H_0.

Another name for the p-value is the "statistical significance of the test". The smaller the p-value, the more statistically significant the test result. We can conclude that the moving average strategy "statistically significantly outperforms" the buy-and-hold strategy if the p-value is low enough to warrant a rejection of H_0. Conventional statistical significance levels are 1%, 5%, and 10%. It is worth mentioning that 1% significance level is a very tough requirement for rejecting the null hypothesis. This means that the chance that the outperformance produced by the moving average strategy is a "false discovery" is less than 1%.

7.2.3 Parametric Tests

A parametric test of hypothesis (7.5) is a test based on the assumption that random variables r_t^e and R_t^e follow a specific probability distribution. Most often, for the sake of simplicity, one assumes that these two random variables follow a bivariate normal distribution. In other words, each of these two random variables follows a normal distribution and, besides, these two random variables are correlated. This type of test is "parametric", because each random variable is assumed to have the same probability distribution that is parameterized by mean and standard deviation.

A parametric hypothesis test is typically specified in terms of a "test statistic". A test statistic is a standardized value that is calculated from sample data. This test statistic follows a well-known distribution and, thus, can be used to calculate the p-value. In our context, because various performance measures are computed differently, each specific performance measure requires using a specific test statistics. The advantage of a parametric test is that one can calculate the p-value of the test fast and quick. Unfortunately, not all performance measures have theoretically computed test statistics. Whereas the mean excess return and the Sharpe ratio have theoretically computed test statistics, the Sortino ratio has not.

Using the mean excess return as a performance measure has some statistical advantages. Specifically, the Central Limit Theorem in statistics says that as long as the excess returns are independent and identically distributed, the mean excess return becomes normally distributed if a sample is large enough.

In this case, given $\hat{\rho}$ as the estimated correlation coefficient between the two series of excess returns, the test of the null hypothesis is performed using the following test statistic

$$z = \frac{\bar{r}_{MA} - \bar{r}_{BH}}{\sqrt{\frac{1}{T}\left(\hat{\sigma}_{MA}^2 - 2\hat{\rho}\hat{\sigma}_{MA}\hat{\sigma}_{BH} + \hat{\sigma}_{BH}^2\right)}} \tag{7.6}$$

which is asymptotically distributed as a standard normal. This test statistic is equivalent to the standard test statistics for testing the difference between two population means in paired samples. Note that when the two excess return series are not correlated (meaning that $\rho = 0$), the test statistics given by Eq. (7.6) reduces to the standard test statistic for testing the difference between two population means in independent samples (see, for example, Snedecor and Cochran 1989).

When the performance is measured by the Sharpe ratio, one can employ the Jobson and Korkie (1981) test with the Memmel (2003) correction. This test assumes the joint normality of the two series of excess returns and is obtained via the test statistic

$$z = \frac{\widehat{SR}_{MA} - \widehat{SR}_{BH}}{\sqrt{\frac{1}{T}\left[2(1 - \hat{\rho}) + \frac{1}{2}(\widehat{SR}_{MA}^2 + \widehat{SR}_{BH}^2 - 2\hat{\rho}^2\widehat{SR}_{MA}\widehat{SR}_{BH})\right]}}, \tag{7.7}$$

which is asymptotically distributed as a standard normal when the sample size is large.

7.2.4 Non-Parametric Tests

Parametric tests are based on a number of assumptions. The standard assumptions are that return distributions are normal and stationary, without serial dependency, and sample sizes are large. Unfortunately, these assumptions are not met in the real world. Specifically, the financial econometrics literature documents that empirical return distributions are non-normal and heteroscedastic (that is, volatility is changing over time); often the series of returns exhibit serial dependence. Consequently, the standard assumptions in parametric tests are generally violated and, therefore, these tests are usually invalid.

Non-parametric tests do not require making assumptions about probability distributions. Most often, non-parametric tests employ computer-intensive randomization methods to estimate the distribution of a test statistic. Non-parametric randomization tests are slower than parametric tests, but have numerous advantages. Besides the fact that they are distribution-free, these methods provide accurate results even if the sample size is small; the test statistic can be chosen freely; the implementation of the test is simple and similar regardless of the choice of a performance measure.

The "bootstrap" is the most popular computer-intensive randomization method that is based on resampling the original data. The practical realization of a bootstrap method depends crucially on whether the time series of excess returns are assumed to be serially independent or dependent. We will refer to the bootstrap method for serially independent data as the "standard bootstrap". The most popular bootstrap methods for serially dependent data are "block bootstraps".

7.2.4.1   Standard Bootstrap

The standard bootstrap method was introduced by Efron (1979). The method is implemented by resampling the data randomly with replacement. In our context, the data are represented by a paired sample of observations of excess returns $\{r_t^e\}$ and $\{R_t^e\}$ where $t = \{1, 2, \ldots, T\}$. This method consists in drawing N random resamples $t^b = \{s_1^b, s_2^b, \ldots, s_T^b\}$, where b is an index for the bootstrap number (so $b = 1$ for bootstrap number 1) and where each of the time indices $s_1^b, s_2^b, \ldots, s_T^b$ is drawn randomly with replacement from $1, 2, \ldots, T$. Each random resample t_b is used to construct the pseudo-time series of the two excess returns $\{r_{t_b}^e\}$ and $\{R_{t_b}^e\}$. Notice that, because the pair $\left(r_{s_i^b}^e, R_{s_i^b}^e\right)$ (where $i \in \{1, T\}$) represents two original excess returns observed at the same time, this method creates two pseudo-time series that retain the historical correlation between the original data series. Observe also that the number of observations in each resample equals the number of observations in the original sample.

For each pseudo-time series of the two excess returns, the difference $\widehat{\Delta}^b$ between the estimated performance measures is computed. By repeating the resampling procedure N times and calculating each time $\widehat{\Delta}^b$, the bootstrap distribution of $\widehat{\Delta}$ is constructed. Finally, to estimate the significance level for the hypothesis test given by (7.5), one can count how many times the computed value of $\widehat{\Delta}^b$ after randomization falls below zero. If the number of negative values of $\widehat{\Delta}^b$ in the bootstrapped distribution is denoted by n, the p-value of the test is computed as $\frac{n}{N}$. It is worth noting that this p-value is computed using a sort of "indirect" test of the null hypothesis. This is because the original data are used that do not satisfy the null hypothesis. If one wanted to carry out a "direct" test, one would have to resample from probability distributions that satisfied the constraint of the null hypothesis, that is, from some modified data where the two empirical performance measure were exactly equal (see a similar discussion in Ledoit and Wolf 2008).

7.2.4.2   Block Bootstrap

The standard bootstrap method assumes that the data are serially independent. If the data are serially dependent, the standard bootstrap cannot be used because it breaks up the serial dependence in the data. That is, it creates serially independent resamples. In order to preserve the dependence structure of the original data series while performing a bootstrap method, one can resample the data using blocks of data instead of individual observations. There are basically two different ways of proceeding, depending on whether the blocks are overlapping or non-overlapping. Carlstein (1986) proposed non-overlapping blocks for univariate time series data, whereas Künsch (1989) suggested overlapping blocks in the same setting. We will refer to the former and the latter method as the non-overlapping block bootstrap method and the moving block bootstrap method respectively. The moving block bootstrap method, considered below, is preferable because it can be used when the sample size is small relative to a block length.[6]

Let l denote the block length. In the moving block bootstrap method, the total number of overlapping blocks in a sample of size T equals $T - l + 1$, where the ith block is given by time indices $B_i = (i, i + 1, \ldots, i + l - 1)$ for $1 \leq i \leq T - l + 1$. Denote by m the number of non-overlapping blocks in the sample (and each random resample) and suppose, for the sake of simplicity of exposition, that $T = m \times l$. In particular, m denotes the required number of blocks that, when placed one after the other, create a sample of size T. The block bootstrap method consists in drawing N random resamples $t^b = \{B_1^b, B_2^b, \ldots, B_m^b\}$ where each block of time indices B_i^b is drawn randomly with replacement from among available blocks $B_1, B_2, \ldots, B_{T-l+1}$. As in the standard bootstrap method, each random resample t_b is used to construct the pseudo-time series of the two excess returns $\{r_{t_b}^e\}$ and $\{R_{t_b}^e\}$. The computation of the p-value of the null hypothesis also goes along similar lines as in the standard bootstrap method.

By construction, in the moving block bootstrap method the bootstrapped time series have a non-stationary (conditional) distribution. The resample becomes stationary if the block length is random. This version of the moving

---

[6]For the sake of illustration, suppose that the sample size is 30 and the block length is 5. In this case, there are only 6 non-overlapping blocks of data in the sample. In contrast, the number of overlapping blocks equals 26.

block bootstrap is called the "stationary bootstrap" and was introduced by Politis and Romano (1994). In particular, unlike the original moving block bootstrap method where the block length is fixed, in the stationary block bootstrap method the length of block B_i^b, l_i^b, is generated from a geometric distribution with probability p. Thus, the average block length equals $\frac{1}{p}$. Therefore p is chosen according to $p = \frac{1}{l}$ where l is the required average block length. The ith block begins from a random index i which is generated from the discrete uniform distribution on $\{1, 2, \ldots, T\}$. Since a generated block length is not limited from above, $l_i^b \in [1, \infty)$, and the block can begin with observation on time T, the stationary bootstrap method "wraps" the data around in a "circle", so that 1 follows T and so on.

The question of paramount importance in the implementation of the block bootstrap method is how to choose the optimal block length. The paper by Hall et al. (1995) addresses this issue. The authors find that the optimal block length depends very much on context. In particular, the asymptotic formula for the optimal block length is $l \sim T^{\frac{1}{h}}$, where $h = 3$, 4, or 5. For computing block bootstrap estimators of variance, $h = 3$. For computing block bootstrap estimators of one-sided and two-sided distribution functions of the test statistic of interest, $h = 4$ and 5 respectively. Since our hypothesis test given by (7.5) corresponds to one-tailed test, then, for example, if the number of observations $T = 1000$, the optimal block length can be roughly estimated as $1000^{\frac{1}{4}} \approx 6$. Another method of selection of the optimal block length is proposed by Politis and White (2004) (see also the subsequent correction of the method by Patton et al. 2009).

7.3 Chapter Summary and Additional Remarks

Simulating the returns to a moving average trading strategy is trivial. In contrast, the question of whether the moving average strategy outperforms its passive counterpart has no unique answer. The literature on portfolio performance measurement starts with the seminal paper of Sharpe (1966) who proposed a reward-to-risk measure now widely known as the Sharpe ratio. However, since the Sharpe ratio uses the standard deviation as a risk measure, it has been often criticized because, apparently, the standard deviation is not able to adequately measure the risk. The literature on performance evaluation, where researchers replace the standard deviation in the Sharpe ratio by an alternative risk measure, is a vast one.

However, there is another stream of research that advocates that the choice of performance measure does not influence the evaluation of risky portfolios. For example, Eling and Schuhmacher (2007), Eling (2008), and Auer (2015) computed the rank correlations between the rankings produced by a set of alternative performance measures (including the Sharpe ratio), and found that the rankings are extremely positively correlated. These researchers concluded that the choice of performance measure is irrelevant, since, judging by the values of rank correlations, all measures give virtually identical rankings. The explanation of this finding is given by Zakamulin (2011) who, among other things, demonstrated analytically that many alternative performance measures produce the same ranking of risky assets as the Sharpe ratio when return distributions are normal. As a result, deviations from normality must be economically significant to warrant using an alternative to the Sharpe ratio.

From a practical point of view, the findings in the aforementioned studies advocate that the choice of a reward-to-risk performance measure is not crucial in testing whether the moving average trading strategy outperforms the buy-and-hold strategy. The Sharpe ratio is the best known and best understood performance measure and, therefore, might be considered preferable to other performance measures from a practitioner's point of view. Yet, one has to keep in mind that the justification of the Sharpe ratio, as well as any other sensible reward-to-risk ratio, depends significantly on the assumptions of existence of the risk-free asset and unrestricted borrowing at the risk-free rate.

It is important to understand that, in order to conclude that the moving average strategy outperforms its passive counterpart, it is not enough to find that the estimated performance measure of the moving average strategy is higher than that of the buy-and-hold strategy. One needs to verify statistically whether the outperformance is genuine or spurious. In other words, the outperformance is reliable only when the estimated performance measure of the moving average strategy is statistically significantly higher than that of the buy-and-hold strategy. To test the outperformance hypothesis, one can use either parametric or non-parametric methods. Parametric methods are fast and simple, but require making a number of assumptions that are usually not satisfied by empirical data. Non-parametric methods are computer-intensive, but require fewer assumptions and more accurate. In testing the outperformance hypothesis, the stationary (block) bootstrap method currently seems to be the preferred method of statistical inference, see, among others, Sullivan et al. (1999), Welch and Goyal (2008), and Kirby and Ostdiek (2012).

References

Auer, B. R. (2015). Does the choice of performance measure influence the evaluation of commodity investments? *International Review of Financial Analysis, 38*, 142–150.

Bodie, Z., Kane, A., & Marcus, A. J. (2007). *Investments*. McGraw Hill.

Carlstein, E. (1986). The use of subseries values for estimating the variance of a general statistic from a stationary sequence. *Annals of Statistics, 14*(3), 1171–1179.

Cogneau, P., & Hübner, G. (2009). The (More Than) 100 ways to measure portfolio performance. *Journal of Performance Measurement, 13*, 56–71.

Efron, B. (1979). Bootstrap methods: Another look at the Jackknife. *Annals of Statistics, 7*(1), 1–26.

Eling, M. (2008). Does the measure matter in the mutual fund industry? *Financial Analysts Journal, 64*(3), 54–66.

Eling, M., & Schuhmacher, F. (2007). Does the choice of performance measure influence the evaluation of hedge funds? *Journal of Banking and Finance, 31*(9), 2632–2647.

Hall, P., Horowitz, J. L., & Jing, B.-Y. (1995). On blocking rules for the bootstrap with dependent data. *Biometrika, 82*(3), 561–574.

Jobson, J. D., & Korkie, B. M. (1981). Performance hypothesis testing with the Sharpe and Treynor measures. *Journal of Finance, 36*(4), 889–908.

Kirby, C., & Ostdiek, B. (2012). It's all in the timing: Simple active portfolio strategies that outperform naive diversification. *Journal of Financial and Quantitative Analysis, 47*(2), 437–467.

Künsch, H. R. (1989). The Jacknife and the Bootstrap for general stationary observations. *Annals of Statistics, 17*(3), 1217–1241.

Ledoit, O., & Wolf, M. (2008). Robust performance hypothesis testing with the Sharpe Ratio. *Journal of Empirical Finance, 15*(5), 850–859.

Levy, H., & Markowitz, H. (1979). Approximating expected utility by a function of mean and variance. *American Economic Review, 69*(3), 308–317.

Memmel, C. (2003). Performance hypothesis testing with the Sharpe Ratio. *Finance Letters, 1*, 21–23.

von Neumann, J., & Morgenstern, O. (1944). *Theory of games and economic behavior.* Princeton University Press.

Patton, A., Politis, D., & White, H. (2009). CORRECTION TO: Automatic block-length selection for the dependent Bootstrap by D. Politis and H. White. *Econometric Reviews, 28*(4), 372–375.

Pedersen, C. S., & Satchell, S. E. (2002). On the foundation of performance measures under asymmetric returns. *Quantitative Finance, 2*(3), 217–223.

Politis, D., & Romano, J. (1994). The stationary bootstrap. *Journal of the American Statistical Association, 89*, 1303–1313.

Politis, D., & White, H. (2004). Automatic block-length selection for the dependent bootstrap. *Econometric Reviews, 23*(1), 53–70.

Sharpe, W. F. (1966). Mutual fund performance. *Journal of Business, 31*(1), 119–138.

Sharpe, W. F. (1994). The Sharpe Ratio. *Journal of Portfolio Management, 21*(1), 49–58.

Sortino, F. A., & Price, L. N. (1994). Performance measurement in a downside risk framework. *Journal of Investing, 3,* 59–65.

Snedecor, G. W., & Cochran, W. G. (1989). *Statistical methods* (8th ed.). Iowa State University Press.

Sullivan, R., Timmermann, A., & White, H. (1999). Data-snooping, technical trading rule performance, and the bootstrap. *Journal of Finance, 54*(5), 1647–1691.

Tobin, J. (1969). Comment on Borch and Feldstein. *Review of Economic Studies, 36*(1), 13–14.

Welch, I., & Goyal, A. (2008). A comprehensive look at the empirical performance of equity premium prediction. *Review of Financial Studies, 21*(4), 1455–1508.

Zakamulin, V. (2011). The performance measure you choose influences the evaluation of hedge funds. *Journal of Performance Measurement, 15*(3), 48–64.

Zakamulin, V. (2014). Portfolio performance evaluation with loss aversion. *Quantitative Finance, 14*(1), 699–710.

8

Testing Profitability of Technical Trading Rules

8.1 Problem Formulation

The ultimate question we want to answer is whether some moving average trading rules outperform the buy-and-hold strategy. If the answer to this question is affirmative, then we want to know the types of rules that perform best. In addition, there are many financial asset classes: stocks, bonds, foreign currencies, real estate, commodities, etc. Therefore the natural additional question to ask is in which markets the moving average trading rules are most profitable?

The difficulty in testing the profitability of moving average trading rules stems from the fact that the procedure of testing involves either a single- or multi-variable optimization. Specifically, any moving average trading rule considered in Chap. 4 has at least one parameter that can take many possible values. For example, in the Moving Average Crossover rule, $MAC(s, l)$, there are two parameters: the size of the shorter averaging window s and the size of the longer averaging window l. As a result, testing this trading rule using relevant historical data consists in evaluating performance of the same rule with many possible combinations of (s, l). When daily data are used, the number of tested combinations can easily exceed 10,000. Besides, there are many types of moving averages (SMA, LMA, EMA, etc.) that can be used in the computation of the average values in the shorter and longer windows. This further increases the number of specific realizations of the same rule that need to be tested. If one additionally considers other types of rules ($MOM(n)$, $\Delta MA(n)$, $MACD(s, l, n)$, etc.) and several data frequencies (daily, weekly, monthly), then one needs to test an exceedingly huge number of specific rules. The main problem in this case, when a great number of specific rules are tested,

© The Author(s) 2017
V. Zakamulin, *Market Timing with Moving Averages*, New Developments
in Quantitative Trading and Investment, DOI 10.1007/978-3-319-60970-6_8

is not computational resources,[1] but how to correctly perform the statistical test of the outperformance hypothesis. Notice that in the preceding chapter we considered how to test the outperformance hypothesis for a single specific rule. Testing the outperformance hypothesis for a trading rule that involves parameter optimization is much more complicated.

In the subsequent sections we review two major types of tests that are used in finance to evaluate the performance of trading rules that require parameter optimization: back-tests (or in-sample tests) and forward tests (or out-of-sample tests). Throughout the exposition, we focus on discussing the advantages and disadvantages of each type of test.

8.2   Back-Testing Trading Rules

In our context, back-testing a trading rule consists in simulating the returns to this trading rule using relevant historical data and checking whether the trading rule outperforms its passive counterpart. However, because each moving average trading rule has at least one parameter, in reality, when a back-test is performed, many specific realizations of the same rule are tested. In the end, the rule with the best observed performance in a back-test is selected and its outperformance is analyzed. This process of finding the best rule among a great number of alternative rules is called "data-mining".

The problem is that the performance of the best rule, found by using the data-mining technique, systematically overstates the genuine performance of the rule. This systematic error in the performance measurement of the best trading rule in a back test is called the "data-mining bias". The reason for the data-mining bias lies in the random nature of returns. Specifically, it is instructive to think about the observed outperformance of a trading rule as comprising two components: the genuine (or true) outperformance and the noise (or randomness):

$$\text{Observed outperformance} = \text{True outperformance} + \text{Randomness.}$$
(8.1)

The random component of the observed outperformance can manifest as either "good luck" or "bad luck". Whereas good luck improves the true outperformance of a trading rule, bad luck deteriorates the true outperformance. It turns out that in the process of data-mining the trader tends to find a rule that benefited most from good luck.

---

[1] In reality, the computational resources are limited. Therefore when a huge number of specific strategies are tested, one can easily stumble upon a lack of computer memory and/or a very slow execution time.

Formal mathematical illustration of the data mining bias is as follows. Suppose that the trader tests a single trading rule by simulating its returns over a relevant historical sample. Suppose that the trader uses either the mean excess return or the Sharpe ratio as performance measure and that the true performance of the trading strategy equals the performance of its passive benchmark. In other words, under our assumption, both strategies perform similarly. In this case the z test statistic (given by either (7.6) or (7.7)) is normally distributed with zero mean and unit variance. The test of a single strategy is not data-mining and, selecting the appropriate significance level α, the p-value of a single test is given by

$$p_S = Prob(z > z_{1-\alpha}), \tag{8.2}$$

where $z_{1-\alpha}$ is the $1 - \alpha$ quantile of the standard normal distribution.[2] If the significance level $\alpha = 0.05$, then $p_S = 5\%$. That is, the probability of "false discovery" amounts to 5% in a single test.

Now suppose that the trader tests N independent strategies and the true performance of each of these strategies equals that of the passive benchmark. That is, we assume that all strategies perform similarly.[3] Under these assumptions the test statistics for these N strategies are independent. Let us compute the probability that with multiple testing at least one of these N strategies produces a p-value below the chosen significance level α. This probability is given by

$$p_N = 1 - Prob(z_1 < z_{1-\alpha}; z_2 < z_{1-\alpha}; \ldots; z_N < z_{1-\alpha}) = 1 - (1 - p_S)^N, \tag{8.3}$$

where $z_i, i \in [1, N]$, is the value of test statistic for strategy i. Notice that this p-value, p_N, is computed as one minus the probability that in all independent tests the p-values are less than α. Since in a single test $Prob(z_i < z_{1-\alpha}) = 1 - p_S$ and all tests are independent, the probability that in N independent tests all p-values are less than α equals $(1 - p_S)^N$.

If in a single test $p_S = 5\%$ and $N = 10$, then $p_N = 40.1\%$. That is, if the trader tests 10 different strategies, then the probability that the trader finds at

---

[2] If $\Phi(x)$ is the cumulative distribution function of a standard normal random variable, the quantile function $\Phi^{-1}(p)$ is the inverse of the cumulative distribution function. The $1 - \alpha$ quantile is given by $z_{1-\alpha} = \Phi^{-1}(1 - \alpha)$. For example, if $\alpha = 5\%$, then $z_{95\%} \approx 1.64$. That is, a standard normal random variable exceeds 1.64 with probability of 5%. Thus, if in a single test the value of test statistics exceeds 1.64, the outperformance delivered by the active strategy is statistically significant at the 5% level.

[3] Note that our example is purely hypothetical where, by assumption, the true performance of all strategies equals the performance of the passive benchmark. The goal of this hypothetical example is to illustrate that in this situation there is a rather high probability that the trader falsely discovers that some strategies statistically significantly outperform the benchmark.

least one strategy that "outperforms" the passive benchmark is about 40%. If $N = 100$, then $p_N = 99.4\%$. It implies the probability of almost 100% that the trader finds at least one strategy that "outperforms" the passive benchmark if the number of tested strategies equals 100. In the context of equation (8.1), for all of the tested strategies in our example the true outperformance equals zero. Consequently, the observed outperformance of each strategy is the result of pure luck (randomness). Therefore the selected best strategy in a back test is the strategy that benefited most from luck.

The data-mining technique is based on a multiple testing procedure which greatly increases the probability of "false discovery" (Type I error in statistical tests). That is, when more than one strategy is tested, false rejections of the null hypothesis of no outperformance are more likely to occur; the trader more often incorrectly "discovers" a profitable trading strategy. To deal with the data-mining bias in multiple back-tests, one has to adjust the p-value of a single test. Since the observed performance of the best rule in a back test is positively biased, to estimate the true outperformance one has to adjust downward the observed performance.[4]

In multiple testing, the usual p-value p_S for a single test no longer reflects the statistical significance of outperformance; the correct statistical significance of outperformance is reflected by p_N. If the test statistics are independent, the adjusted p-value of a single test can be obtained by

$$p_S^* = 1 - (1 - p_N)^{\frac{1}{N}}. \tag{8.4}$$

For example, if $N = 10$ strategies are tested and $p_N = 5\%$, to reject the null hypothesis that a trading strategy does not outperform its passive counterpart, the p-value of the test statistic in a single test must be below 0.5%. If 100 strategies are tested, the p-value of a single test must be below 0.05%. However, in reality the returns to tested strategies are correlated. As a result, their test statistics are dependent and the adjustment method must take into account their dependence.[5] Different methods of performing a correct statistical inference in multiple back-tests of trading rules are discussed in Markowitz and Xu (1994), White (2000), Hansen (2005), Romano and Wolf (2005), Harvey and Liu (2014), Bailey and López de Prado (2014), and Harvey and Liu (2015).

---

[4]For example, a common practice in evaluating the true performance of the best rule in a back test is to discount the reported Sharpe ratio by 50%, see Harvey and Liu (2015).

[5]If the test statistics are perfectly correlated, then the p-value in a multiple test equals the p-value in a single test. Consequently, when the test statistics are neither independent nor perfectly correlated and the p-value of a single test is given by p_S, the adjusted p-value lies somewhere in between p_S and $1 - (1 - p_S)^{\frac{1}{N}}$.

The main advantage of back-tests is that they utilize the full historical data sample. Since the longer the sample the larger the power of any statistical test, back-tests decrease the chance of missing "true" discoveries, that is, the chance of missing profitable trading strategies. However, because all methods of adjusting p-values in multiple tests try to minimize a Type I error in statistical tests (probability of false discovery), this adjustment also greatly increases the probability of missing true discoveries (Type II error in statistical tests). As an example, suppose that a trading strategy truly outperforms its passive benchmark and the p-value of a single test is 0.1%. If the 5% significance level is used, the outperformance of this strategy is highly statistically significant if not outstanding. However, if, in addition to this strategy, the trader tests another 99 trading strategies, the trader has to use 0.05% significance level in a single test (assuming that all test statistics are independent). As a result, in a multiple test the outperformance of this strategy is no longer statistically significant. The trader fails to detect this strategy with genuine outperformance, because this strategy simply had a bad luck to be a part of a multiple test.

As final remarks regarding the back-tests and data-mining bias, it is worth mentioning the following. The data-mining bias decreases when the sample size increases. This is because the larger the sample size, the lesser the effect of randomness in the estimation of performances of trading rules.[6] The data-mining bias increases with increasing number of rules. Adding a new tested rule to the existing set of tested rules cannot decrease the performance of the best rule in a back test. In particular, if the new rule performs worse than the best rule among the existing set of rules, the performance of the best rule in a back test remains the same. If, on the other hand, the new rule performs better, then the new rule becomes the best performing rule.

8.3   Forward-Testing Trading Rules

To mitigate the data-mining bias problem in back-testing trading rules, instead of adjusting the p-value and/or performance of the best rule, an alternative solution is to perform forward testing trading rules. The idea behind a forward test is pretty straightforward: since the performance of the best rule in a back test overstates the genuine performance of the rule, to validate the rule and to provide an unbiased estimate of its performance, the rule must be tested using an additional sample of data (besides the sample used for back-testing the rules). In other words, a forward test augments a back test with an additional validation test. For this purpose the total sample of historical data is segmented into a

---

[6]However, this property holds true only when the market's dynamics is not changing over time.

"training" set of data and a "validation" set of data. Most often, the training set of data that is used for data-mining is called the "in-sample" segment of data, while the validation set of data is termed the "out-of-sample" segment. In this regard, the back-tests are often called the "in-sample" tests, whereas the forward tests are called the "out-of-sample" tests.

To understand the forward testing procedure, suppose that the trader wants to forward test the performance of the Momentum rule $MOM(n)$. The forward testing procedure begins with splitting the full historical data sample $[1, T]$ into the in-sample subset $[1, s]$ and out-of-sample subset $[s + 1, T]$, where T is the last observation in the full sample and s denotes the split point. Then, using the training set of data, the trader determines the best window size n^* to use in this rule. Formally, the choice of the optimal n^* is given by

$$n^* = \arg \max_{n \in [n^{\min}, n^{\max}]} \mathcal{M}(r^e_{1,n}, r^e_{2,n}, \ldots, r^e_{s,n}),$$

where $n^{\min}$ and $n^{\max}$ are the minimum and maximum values for n respectively, $\mathcal{M}$ is the performance measure preferred by the trader, and $(r^e_{1,n}, r^e_{2,n}, \ldots, r^e_{s,n})$ are the excess returns to the Momentum rule with window size n over the training dataset $[1, s]$. Finally, the best rule discovered in the mined data (in-sample) is evaluated on the out-of-sample data. That is, the trader evaluates the out-of-sample performance of the $MOM(n^*)$ rule

$$\text{Out-of-sample performance} = \mathcal{M}(r^e_{s+1,n^*}, r^e_{s+2,n^*}, \ldots, r^e_{T,n^*}),$$

where $(r^e_{s+1,n^*}, r^e_{s+2,n^*}, \ldots, r^e_{T,n^*})$ are the excess returns to the Momentum rule with window size n^* over the out-of-sample set of data $[s + 1, T]$.

In practical implementations of out-of-sample tests, the in-sample segment of data is usually changed during the test procedure. Depending on the assumption of whether or not the market's dynamics is changing over time, either expanding or rolling in-sample window is used. If the market's dynamics is stable, the best trading rule is not changing over time. Therefore, after a period of length $p \geq 1$, at time $s + p$, the trader can repeat the best trading rule selection procedure using a longer in-sample window $[1, s + p]$. Afterwards, the procedure of selecting the best trading rule can be repeated at times $s + 2p$, $s + 3p$, and so on. Notice that, since the in-sample segment of data always starts with observation number 1, the size of the in-sample window increases with each iteration of the selection of best rule procedure. The rationale behind using an expanding window in out-of-sample tests is the notion that the longer the sample of data, the smaller the data-mining bias and, therefore, the better precision in identifying the best trading rule.

Observe the following sequence of steps in the out-of-sample testing procedure with expanding in-sample window. First, the best parameters are estimated using the in-sample window $[1, s]$ and the returns to the best rule are simulated over the out-of-sample period $[s + 1, s + p]$. Next, the best parameters are re-estimated using the in-sample window $[1, s + p]$ and the returns to the new best rule are simulated over the out-of-sample period $[s + p + 1, s + 2p]$. This sequence of steps is repeated until the returns are simulated over the whole out-of-sample period $[s + 1, T]$. In the end, the trader evaluates the performance of the trading strategy over the whole out-of-sample period.

However, if the trader believes that the market's dynamics is changing over time, the use of the expanding window is no longer optimal. Instead, a rolling in-sample window must be used. The technique of using a rolling in-sample window in a forward test is usually called a walk-forward test (or out-of-sample test with a rolling/moving window). Specifically, after the initial determination of the best trading rule over the data segment $[1, s]$, the trader simulates the returns to the trading rule over $[s + 1, s + p]$, and then repeats the procedure of selecting the best trading rule using a new in-sample window $[1 + p, s + p]$. Notice that in this case the length of the in-sample window always equals s, but with each iteration of the selection of best rule procedure, the in-sample window is moved forward by step size p. The premise behind using a rolling window in out-of-sample tests is the notion that, when market's dynamics is changing, the recent past is a better foundation for selecting trading rule parameters than the distant past.

Figure 8.1 provides illustrations of the out-of-sample testing procedure with an expanding in-sample window and a rolling in-sample window. It is worth noting that the out-of-sample testing methodology with either an expanding or rolling window has a dynamic aspect, in which the trading rule is being modified over time as the market evolves. The out-of-sample methodology

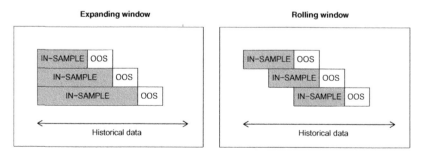

Fig. 8.1 Illustration of the out-of-sample testing procedure with an expanding in-sample window (*left panel*) and a rolling in-sample window (*right panel*). OOS denotes the out-of-sample segment of data for each in-sample segment

closely resembles the real-life trading where a trader, at each point in time, has to make a choice of what trading rule to use given the information about the past performances of many trading rules. Therefore an out-of-sample test is not a test of whether some specific trading rule outperforms the passive benchmark, but rather a test of whether the trader can beat the benchmark by using a set of various rules and, at any time, following a strategy with the best observed performance in a back test.

The great advantage of out-of-sample testing methods is that they, at least in theory, should provide an unbiased estimate of the rule's true outperformance. An additional advantage is that the out-of-sample simulation of returns to a trading strategy, with subsequent measurement of its performance, are relatively easy to do as compared to the implementation of rather sophisticated performance adjustment methods in multiple back-tests. However, out-of-sample testing methods have one unresolved deficiency that may seriously corrupt the estimation of the true outperformance of a trading rule. The primary concern is that no guidance exists on how to choose the split point between the in-sample and out-of-sample subsets. One possible approach is to choose the initial in-sample segment with a minimum length and use the remaining part of the sample for the out-of-sample test (see Marcellino et al. 2006, and Pesaran et al. 2011). Another potential approach is to do the opposite and reserve a small fraction of the sample for the out-of-sample period (as in Sullivan et al. 1999). Alternatively, the split point can be selected to lie somewhere in the middle of the sample. The problem is that when the in-sample segment is short, the data-mining bias is substantial and researchers increase the chance of making "false" discoveries. On the other hand, when the out-of-sample segment is short, the statistical power of outperformance tests is reduced and researchers increase the chance of not rejecting a false null hypothesis of no outperformance. In any case, regardless of the choice of split point, the conventional wisdom says that the out-of-sample performance of a trading strategy provides an unbiased estimate of its real-life performance.

Yet recently the conventional wisdom about the unbiased nature of traditional out-of-sample testing has been challenged. In the context of out-of-sample forecast evaluation, Rossi and Inoue (2012) and Hansen and Timmermann (2013) report that the results of out-of-sample forecast tests depend significantly on how the sample split point is determined. In the context of out-of-sample performance evaluation of trading rules, Zakamulin (2014) also demonstrates that the out-of-sample performance of trend following strategies depends critically on the choice of the split point. The primary argument (put forward in the paper by Zakamulin 2014), for why the choice of the split point sometimes dramatically affects the out-of-sample performance of a trend following strategy, lies in the fact that the outperformance delivered by any trend following strategy is highly non-uniform. Generally, a trend following strategy underperforms its passive benchmark during bull markets and shows a superior performance during bear markets. This argument means that the choice of the split point cannot be made arbitrary: researchers must ensure that both the in-sample and out-of-sample segments contain alternating bull and bear market periods (alternating periods of upward and downward trends). A failure of not including both bull and bear periods into the in-sample segment of data results in selecting a trading rule that is not "trained" to detect changes in trends. Similarly, a failure of not including bear periods into the out-of-sample segment of data results in erroneous rejection of "true" discoveries.[7]

Last but not least, the findings reported by Zakamulin (2014) also mean that the traditional out-of-sample tests are not free from "data-mining" issues. Specifically, using real historical data, Zakamulin provides an example where he demonstrates that, depending on the choice of the split point, the out-of-sample performance of a trend following rule might be either superior or inferior as compared to that of its passive counterpart. Therefore, in principle, a researcher might consider multiple split points and report the out-of-sample performance that most favors a trading rule.

8.4   Chapter Summary and Additional Remarks

Each moving average trading rule considered in this book has from one to three parameters which values are not pre-specified. In practical applications of these rules, the trader has to make a choice of which specific parameters to use. Therefore, traders inevitably tend to search over a large number of parameters in the attempt to optimize a trading strategy performance using relevant historical data. This procedure of selecting the best parameters to use in a trading rule is called back-testing.

However, financial researchers long ago realized that when a large number of technical trading rules are searched, this selection procedure tends to find a rule which performance benefited most from luck (see, for example, Jensen 1967). Therefore the observed performance of the best rule in a back-test tends

---

[7]For example, Sullivan et al. (1999) use the period from 1987 to 1996 for out-of-sample tests and find that no any technical trading rule outperforms the passive benchmark in out-of-sample test. However, this whole out-of-sample historical period can be considered as a single long bull market. That is, virtually during the whole out-of-sample period the stock prices trended upward. Since the outperformance delivered by trend following rules appears as a result of protection from losses during bear markets, no wonder that these researchers found that during a bull market no any technical trading rule outperforms.

to systematically overestimate the rule's genuine performance.[8] This systematic error is called the data-mining bias. The methods of correcting the data-mining bias appeared relatively recently; probably the first published paper on this topic was the paper by Markowitz and Xu (1994). Unfortunately, one can still find recently published papers in scientific journals where researchers employ back-tests and document the observed performance of the best rules without correcting for data-mining bias.

Besides the data-mining correction methods that adjust downward the performance of the best trading rule in a back test, the other straightforward method of the estimation of true performance of a trading rule is to employ a validation procedure. The method of combination of a back-test with a subsequent validation test is called a forward-test and was proposed already by Jensen (1967). Another name for this test is an out-of-sample test. The application of forward-testing trading rules started already in early 1970s. The first applications of the so-called "walk-forward tests" that use a rolling training window can be found in the papers by Lukac et al. (1988) and Lukac and Brorsen (1990). Surprisingly, in the majority of studies that employ forward-tests of trading rules, the researchers used either the commodity price data or exchange rate data (see the review paper by Park and Irwin 2007). To the best knowledge of the author, there are only two papers to date in which the researchers implement forward tests of profitability of trading rules in stock markets: Sullivan et al. (1999) and Zakamulin (2014).

As compared to pure back-tests, forward-tests with either expanding or rolling in-sample window allow a trader to improve significantly the estimation of true performance of trading rules and these procedures closely resemble actual trader behavior. However, forward-tests are not completely superior to back-tests in every respect. Since back-tests make use of the total sample of data, the probability of missing a strategy with genuine outperformance is less than in forward-tests. Forward-tests are supposed to be purely objective out-of-sample tests with no data-mining bias, but in reality they may not be truly out-of-sample. One possibility to corrupt the validity of a forward-test is to try different split points (between the training and validation sets) and report the results that favor most a trading strategy. Another possibility is to try different starting points for a historical sample and choose the starting point that favors most a trading strategy. Yet another possibility is to try many different strategies and report the results only for those strategies that pass the out-of-sample test.

The data-mining problem is, in fact, a part of a larger "data-snooping" problem. As defined in the paper by White (2000) "Data-snooping occurs when a given set of data is used more than once for purposes of inference or

---

[8] Aronson (2006) explains in simple and plain language the cause of the data-mining bias.

model selection". The data-mining procedure in back-tests explicitly re-uses the data many times in searching for the best performing rule. The notion of "data-snooping" also covers the cases where researchers use, either explicitly or implicitly, the results of prior studies of performances of trading rules reported by other researchers. For example, in the studies by Brock et al. (1992), and Siegel (2002, Chap. 17) the authors test the performance of the 200-day SMA rule using the historical prices of the Dow Jones Industrial Average (DJIA) index starting from the index inception in 1896. The authors acknowledge that they test this rule because "it is one of the most popular trading rules among practitioners". It is quite natural to suppose that prior to these studies practitioners back-tested many n-day SMA rules and the 200-day SMA rule was selected as the rule with the best observed performance. In fact, the superior performance of this rule was already documented by Gartley (1935) who also used the prices of the DJIA index. Consequently, one can reasonably suspect that the reported performance of the SMA rule (in the studies by Brock et al. 1992, and Siegel 2002) might be highly overstated as compared to its genuine performance. Unfortunately, it is very difficult to fully avoid the data-snooping problem in empirical studies. To fully avoid this problem requires either using a completely new set of rules or using historical data that do not overlap with the data used in previous studies.

Last but not least, the market's dynamics can change over time. As a result, a profitable rule in the past may not perform well in the future. Even if the rule shows a superior performance in the past, the trader has to examine the consistency of the rule performance over time. That is, the trader has to check whether or not the outperformance deteriorates as time goes. There are way too many examples when the superior performance of a trading rule is confined to a single relatively short particular historical episode.

References

Aronson, D. (2006). *Evidence-based technical analysis: Applying the scientific method and statistical inference to trading signals.* Wiley.

Bailey, D. H., & López de Prado, M. (2014). The deflated Sharpe Ratio: Correcting for selection bias, backtest overfitting, and non-normality. *Journal of Portfolio Management, 40*(5), 94–107.

Brock, W., Lakonishok, J., & LeBaron, B. (1992). Simple technical trading rules and the stochastic properties of stock returns. *Journal of Finance, 47*(5), 1731–1764.

Gartley, H. M. (1935). *Profits in the stock market.* Lambert Gann.

Hansen, P. R. (2005). A test for superior predictive ability. *Journal of Business and Economic Statistics, 23*(4), 365–380.

Hansen, P. R., & Timmermann, A. (2013). *Choice of sample split in out-of-sample forecast evaluation* (Working Paper, European University Institute, Stanford University and CREATES).

Harvey, C. R., & Liu, Y. (2014). Evaluating trading strategies. *Journal of Portfolio Management, 40*(5), 108–118.

Harvey, C. R., & Liu, Y. (2015). Backtesting. *Journal of Portfolio Management, 42*(1), 13–28.

Jensen, M. (1967). Random walks: Reality or myth: Comment. *Financial Analysts Journal, 23*(6), 77–85.

Lukac, L. P., Brorsen, B. W., & Irwin, S. H. (1988). A test of futures market disequilibrium using twelve different technical trading systems. *Applied Economics, 20*(5), 623–639.

Lukac, L. P., & Brorsen, B. W. (1990). A comprehensive test of futures market disequilibrium. *Financial Review, 25*(4), 593–622.

Marcellino, M., Stock, J. H., & Watson, M. W. (2006). A comparison of direct and iterated multistep AR methods for forecasting macroeconomic time series. *Journal of Econometrics, 135*(1–2), 499–526.

Markowitz, H. M., & Xu, G. L. (1994). Data mining corrections. *Journal of Portfolio Management, 21*(1), 60–69.

Park, C.-H., & Irwin, S. H. (2007). What do we know about the profitability of technical analysis? *Journal of Economic Surveys, 21*(4), 786–826.

Pesaran, M. H., Pick, A., & Timmermann, A. (2011). Variable selection, estimation and inference for multi-period forecasting problems. *Journal of Econometrics, 164*(1), 173–187.

Romano, J., & Wolf, M. (2005). Stepwise multiple testing as formalized data snooping. *Econometrica, 73*(4), 1237–1282.

Rossi, B., & Inoue, A. (2012). Out-of-sample forecast tests robust to the choice of window size. *Journal of Business and Economic Statistics, 30*(3), 432–453.

Siegel, J. (2002). *Stocks for the long run.* McGraw-Hill Companies.

Sullivan, R., Timmermann, A., & White, H. (1999). Data-snooping, technical trading rule performance, and the bootstrap. *Journal of Finance, 54*(5), 1647–1691.

White, H. (2000). A reality check for data snooping. *Econometrica, 68*(5), 1097–1126.

Zakamulin, V. (2014). The real-life performance of market timing with moving average and time-series momentum rules. *Journal of Asset Management, 15*(4), 261–278.

Part IV

Case Studies

9

Trading the Standard and Poor's Composite Index

9.1 Data

The Standard and Poor's (S&P) 500 index is a value-weighted stock index based on the market capitalizations of 500 large companies in the US. This index was introduced in 1957 and intended to be a representative sample of leading companies in leading industries within the US economy. Stocks in the index are chosen for market size, liquidity, and industry group representation. This index is probably the most commonly followed equity index and many consider it one of the best representations of the US stock market. The S&P 500 index appeared as a result of expansion of the Standard and Poor's Composite index that was introduced in 1926 and consisted of 90 stocks only. It is common to extend the Standard and Poor's Composite index back in time using the data on the early stock price indices (examples are Shiller 1989, Campbell and Shiller 1998, and Shiller 2000). However, it is worth noting that the data prior to 1926 are less reliable because they are composed from various sources and because of the scarcity of stocks relative to post-1926 data. Practically all of these stocks belong to only two industry groups: "railroad" and "bank and insurance". Schwert (1990) constructed and made publicly available data on the monthly return series for the US stock market beginning from 1802.

Unfortunately, the data for the risk-free rate of return are available from 1857 only. Therefore our full historical sample of monthly data covers the period from January 1857 to December 2015 (159 full years). Nevertheless, our dataset is the longest dataset used for testing moving average trading rules. It should be noted, however, that while long history can provide us with rich information about the past performance of moving average trading rules, the availability of long-term data is both a blessing and a curse. This is because

© The Author(s) 2017   **143**
V. Zakamulin, *Market Timing with Moving Averages*, New Developments
in Quantitative Trading and Investment, DOI 10.1007/978-3-319-60970-6_9

in order to use the observed performance over a very long-term as a reliable estimate of the expected performance in the future, we need to make sure that the stock market dynamics both in the distant and near past were the same. For this purpose we perform a series of robustness tests and tests for regime shifts in the stock market dynamics.

9.1.1   Data Sources and Data Construction

In our empirical study we use the capital gain and total returns (denoted by CAP and TOT respectively) on the Standard and Poor's Composite stock price index, as well as the risk-free rate of return (denoted by RF) proxied by the Treasury Bill rate. Our sample period begins in January 1857 and ends in December 2015, giving a total of 1896 monthly observations. The data on the S&P Composite index come from two sources. The returns for the period January 1857 to December 1925 are provided by William Schwert.[1] The returns for the period January 1926 to December 2015 are computed from the closing monthly prices of the S&P Composite index and corresponding dividend data provided by Amit Goyal.[2] Specifically, the capital gain return is computed using the closing monthly prices, whereas the total return is computed as the sum of the capital gain return and the dividend return.

The Treasury Bill rate for the period January 1920 to December 2015 is also provided by Amit Goyal. Because there was no risk-free short-term debt prior to the 1920s, we estimate it in the same manner as in Welch and Goyal (2008) using the monthly data for the Commercial Paper rates for New York. These data are available for the period January 1857 to December 1971 from the National Bureau of Economic Research (NBER) Macrohistory database.[3] First, we run a regression

$$\text{Treasury Bill rate}_t = \alpha + \beta \times \text{Commercial Paper rate}_t + e_t$$

over the period from January 1920 to December 1971. The estimated regression coefficients are $\alpha = -0.00039$ and $\beta = 0.9156$; the goodness of fit, as measured by the regression R-squared, amounts to 95.7%. Then the values of the Treasury Bill rate over the period January 1857 to December 1919 are obtained using the regression above with the estimated coefficients for the period 1920 to 1971.

---

[1] http://schwert.ssb.rochester.edu/data.htm.

[2] Downloaded from http://www.hec.unil.ch/agoyal/. These data were used in the widely cited paper by Welch and Goyal (2008).

[3] http://research.stlouisfed.org/fred2/series/M13002US35620M156NNBR.

9.1.2 Descriptive Statistics and Evidence for a Regime Shift

Table 9.1 summarizes the descriptive statistics for the monthly returns on the S&P Composite index and the risk-free rate of return. The descriptive statistics are reported for the total historical period from January 1857 to December 2015 as well as for the first and second sub-periods: from January 1857 to December 1943 and from January 1944 to December 2015 respectively. The total sample period spans 159 years, the first and the second sub-periods span 87 and 72 years respectively. The choice of the split point between the sub-periods is motivated by the analysis of a structural break in the growth rate of the index (see below).

The results of the Shapiro-Wilk test reject the normality in all data series over the total period as well as over each sub-period. It is worth noting that over the first sub-period the stock market was much more turbulent than over the second one. In particular, the volatility, as well as the kurtosis, during the first sub-period was considerably higher than that during the second sub-period. On the other hand, over the first sub-period the capital gain returns and total returns were substantially lower than those over the second sub-period. In addition, over the first sub-period the stock return series exhibited a statistically and economically significant positive autocorrelation. In contrast,

Table 9.1 Descriptive statistics for the monthly returns on the S&P Composite index and the risk-free rate of return

	1857–2015			1857–1943			1944–2015		
Statistics	CAP	TOT	RF	CAP	TOT	RF	CAP	TOT	RF
Mean, %	5.88	10.29	3.88	3.94	9.13	3.75	8.23	11.69	4.04
Std. dev., %	17.50	17.51	0.75	19.72	19.72	0.60	14.36	14.40	0.90
Skewness	0.13	0.18	0.98	0.34	0.38	0.67	−0.42	−0.41	0.93
Kurtosis	7.99	8.20	2.60	8.46	8.68	4.45	1.58	1.60	1.06
Shapiro-Wilk	**0.93**	**0.93**	**0.93**	**0.91**	**0.91**	**0.93**	**0.98**	**0.98**	**0.93**
	(0.00)	(0.00)	(0.00)	(0.00)	(0.00)	(0.00)	(0.00)	(0.00)	(0.00)
AC_1	**0.07**	**0.07**	**0.97**	**0.09**	**0.09**	**0.94**	0.03	0.03	**0.99**
	(0.00)	(0.00)	(0.00)	(0.00)	(0.00)	(0.00)	(0.45)	(0.41)	(0.00)

Notes CAP, TOT, and RF denote the capital gain return, the total market return, and the risk-free rate of return respectively. Means and standard deviation are annualized and reported in percentages. **Shapiro-Wilk** denotes the value of the test statistics in the Shapiro-Wilk normality test. The p-values of the normality test are reported in brackets below the test statistics. AC_1 denotes the first-order autocorrelation. For each AC_1 we test the hypothesis $H_0 : AC_1 = 0$. The p-values are reported in brackets below the values of autocorrelation. Bold text indicates values that are statistically significant at the 5% level

over the second sub-period the autocorrelation in stock returns was neither economically nor statistically significant.

The results reported in Table 9.1 suggest that the stock market mean (capital gain and total) returns and volatilities were different across the two sub-periods. To find out whether these differences are statistically significant, we perform the tests of the stability of mean returns and variances across the two sub-periods. The results of these tests suggest that there is strong statistical evidence that the variances of both the capital gain and total returns have changed over time (see the subsequent appendix for the detailed description of the tests and their results). Besides, whereas we cannot reject the hypothesis that the mean total market return has been stable over time, we can reject the hypothesis about the stability of the mean capital gain return over time.

Thereby our results advocate that there are economically and statistically significant differences in the mean capital gain returns across the two sub-periods of data. In particular, over the first sub-period the mean annual capital gain return was about 4%, whereas over the second sub-period the mean annual capital gain return was approximately 8%. Consequently, over the second sub-period the mean capital gain return on the S&P Composite index was double as much as that over the first sub-period. Since the trading signal in all moving average rules is computed using closing prices not adjusted for dividends,[4] this finding may be of paramount importance for testing the performance of trading rules.

In order to verify that there is a major break in the growth rate of the S&P Composite index, we perform an additional structural break analysis. The goal of this analysis is to test for the presence of a single structural break in the growth rate of the S&P Composite index. The null hypothesis in this test is that the period t log capital gain return on the S&P Composite index, r_t, is normally distributed with constant mean μ and variance σ^2. More formally, $r_t \sim \mathcal{N}\left(\mu, \sigma^2\right)$. Under this hypothesis the log of the S&P Composite index at time t is given by the following linear model

$$\log\left(I_t\right) = \log\left(I_0\right) + \sum_{i=1}^{t} r_i = \log\left(I_0\right) + \mu\, t + \varepsilon_t,$$

---

[4]In some published papers the trading signal is computed using the dividend-adjusted prices. However, using dividend-adjusted prices is highly non-standard in traditional technical analysis. In particular, we have studied many handbooks on the technical analysis of financial markets, beginning with the book by Gartley (1935), and in every handbook a technical indicator is supposed to be computed using prices that are easily observable in the market, in contrast to dividend-adjusted prices. Therefore in this book we stick to the standard computation of trading signals. To be on the safe side, we replicated the analysis of the profitability of moving average trading rules using the dividend-adjusted prices. The results of these tests were qualitatively the same as those reported in this chapter. That is, replacing the prices not adjusted by dividends with dividend-adjusted prices does not influence the conclusions reached in our study.

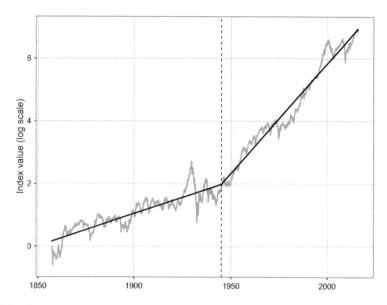

Fig. 9.1 The log of the S&P Composite index over 1857–2015 (*gray line*) versus the fitted segmented model (*black line*) given by $\log(I_t) = \log(I_0) + \mu t + \delta(t - t^*)^+ + \varepsilon_t$, where t^* is the breakpoint date, μ is the growth rate before the breakpoint, and $\mu + \delta$ is the growth rate after the breakpoint. The estimated breakpoint date is September 1944

where I_0 is the index value at time 0 and $\varepsilon_t \sim \mathcal{N}\left(0, \sigma^2 t\right)$. Our alternative hypothesis is that the mean log return at time t^* changes from μ to $\mu + \delta$. Under the alternative hypothesis the log of the S&P Composite index at time t is given by the following segmented model

$$\log(I_t) = \log(I_0) + \mu t + \delta\left(t - t^*\right)^+ + \varepsilon_t,$$

where $(t - t^*)^+$ denotes the positive part of the difference $(t - t^*)$.

The results of the structural break analysis reveal a strong evidence of the presence of a major break in the growth rate of the S&P Composite index around year 1944 (see the subsequent appendix for the detailed description of the structural break analysis and its results). For the sake of illustration, Fig. 9.1 plots the log of the S&P Composite index versus the fitted segmented model.

Given the strong evidence on the occurrence of a major break in the growth rate of the S&P Composite index around 1944, it is natural to ask the following question. What caused this break? In other words, why the price of the S&P Composite index has been growing much faster over the post-1944 period than over the 87-year long period prior to 1944? As a matter of fact, the answer to this question can be readily found in the book by the legendary Benjamin Graham (1949). In particular, in the first part of his book Graham compares the investor's situation in the early 1910s and late 1940s. Regarding the investment practice in the early 1910s, Graham writes that most individual investors bought exclusively high-quality corporate bonds that provided an annual return of about 5%; the income from corporate bonds was fully tax-exempt at that time. He commented further that "There was admittedly such a thing as investment in common stocks; but for the ordinary investor it was either taboo or practiced on a small scale and restricted to a limited number of choice issues". When investors did select some common stocks to invest in, they preferred stocks that provided high and stable dividend income.

The investor's situation underwent dramatic changes from the early 1910s to the late 1940s (and also thereafter). Specifically:

- The rate of return on corporate bonds decreased dramatically as the result of the US government needs to finance World War II. Specifically, the US government faced with the need to raise funds far in excess of tax receipts in order to finance the war effort. To make borrowing cheap, during World War II and thereafter, the Federal Reserve pledged to keep the interest rate on Treasury Bills fixed at 0.375%. The rate of return on corporate bonds fell to about 2.5%. The fixed income market in the US was deregulated only in 1951, see Walsh (1993).
- The inflation increased dramatically as the result of deficit financing during World War I and World War II. In particular, from 1913 to 1920 the average annualized inflation rate was 11%, whereas from 1941 to 1948 the average annualized inflation rate was 7%.[5] Consequently, the rate of return on bonds was way below the inflation rate.
- The interest on bonds, stock dividends, and capital gains became subject to income tax. Specifically, the US government imposed taxes on bond income and capital gains from 1914. Stock dividends were subject to income tax from 1936 to 1939 and from 1953 and thereafter. Income on bonds and stock dividends was taxed at an individual's income tax rate; the top marginal tax rate from 1935 to 1981 was at least 70%. In contrast, capital gains on stocks held for more than six months were taxed at one-half, or less, of the rate applicable to interest and dividends.

As a result of all these changes, investing in corporate bonds no longer made sense for the ordinary individual investors; the return on bonds was far below the inflation rate and this return was heavily taxed. The investors were attracted to common stocks because of a fear of inflation and tax considerations. That is, common stocks seemed to be a natural hedge against inflation. Because of tax

---

[5]These data are available online, see https://www.measuringworth.com/.

considerations, the investors began to prefer stocks with greater capital gains at the expense of dividend income. Buying high paying dividend common stocks no longer made sense for wealthy investors; over 1940s and 1950s the top marginal tax rate increased to about 90%. Therefore high paying dividend stocks went out of favor, and stayed out of favor, beginning from the late 1930s. As a result, firms started to gradually reduce the amount of dividends; dividend payment was gradually replaced by share repurchase.

There is another factor (besides higher demand for stocks) that may provide an additional explanation for why the return on the S&P Composite index increased beginning from the late 1940s. Specifically, the US government substantially increased corporate taxes over 1930s and 1940s. Because corporations pay taxes on their profits after interest payment is deducted, interest expense reduces the amount of corporate tax firms must pay. This interest tax deduction creates an incentive to use debt. Because of this incentive, the debt of US corporations increased dramatically starting from the early 1940s. The gain to investors from the tax deductibility of interest payments is commonly referred to as the "interest tax shield" (see any textbook on corporate finance, for example, Berk and DeMarzo 2013). By increasing the debt, a firm increases its leverage.[6] With or without corporate tax, leverage increases the return on equity.[7]

Everything considered, during 1940s the investors were attracted to stocks because of the low rate of return on bonds, fear of inflation, and tax considerations. Since capital gains were taxed at a much lower rate than dividends, the investors were reluctant to buy stocks with high dividend yield; they preferred buying stocks with large potential capital gains. Therefore beginning from the early 1940s the dividend yield on the S&P Composite index decreased considerably. Since the total return on the index increased while dividend yield decreased, the capital gain return on the S&P Composite index increased substantially in the post-World War II era.

---

[6]The "debt-to-equity" ratio is a common ratio used to assess a firm's leverage.

[7]For the explanation see, for example, Berk and DeMarzo (2013), Chap. 14. In brief, the return on levered equity, r_E is given by $r_E = r_U + \frac{D}{E}(r_U - r_D)$, where r_U is the return on un-levered equity, r_D is the interest on debt, and $\frac{D}{E}$ is the debt-to-equity ratio. For example, if $r_U = 8\%$ and there is no debt, then the return on equity equals 8%. However, if the interest on debt is 4% and the debt-to-equity ratio equals unity (that is, a firm is financed by equity and debt in equal proportions), then the return on equity increases to 12%.

9.2 Bull and Bear Market Cycles and Their Dynamics

It is an old tradition to describe cycles in stock prices as bull and bear markets. Unfortunately, there is no generally accepted formal definition of bull and bear markets in finance literature. There is a common consensus among financial analysts that a bull (bear) market denotes a period of generally rising (falling) prices. However, when it comes to the dating of bull and bear markets, financial analysts are broken up into two distinct groups. One group insists that in order to qualify for a bull (bear) market phase, the stock market price should increase (decrease) substantially. For example, the rise (fall) in the stock market price should be greater than 20% from the previous local trough (peak) in order to qualify for being a distinct bull (bear) market. The other group believes that in order to qualify for a bull (bear) name, the stock market price should increase (decrease) over a substantial period of time. For instance, the stock market price should rise (fall) over a period of greater than 5 months in order to qualify for being a distinct bull (bear) market.

Since there is no unique definition of bull and bear markets, there is no single preferred method to identify the state of the stock market. In our study, to detect the turning points between the bull and bear markets, we employ the dating algorithm proposed by Pagan and Sossounov (2003). This algorithm adopts, with slight modifications, the formal dating method used to identify turning points in the business cycle (Bry and Boschan 1971). The algorithm is based on a complex set of rules and consists of two main steps: determination of initial turning points in raw data and censoring operations. In order to determine the initial turning points, first of all one uses a window of length $\tau_{window} = 8$ months on either side of the date and identifies a peak (trough) as a point higher (lower) than other points in the window. Second, one enforces the alternation of turning points by selecting highest of multiple peaks and lowest of multiple troughs. Censoring operations require: eliminating peaks and troughs in the first and last $\tau_{censor} = 6$ months; eliminating cycles[8] that last less than $\tau_{cycle} = 16$ months; and eliminating phases that last less than $\tau_{phase} = 4$ months (unless the absolute price change in a month exceeds $\theta = 20\%$).[9]

---

[8]A cycle denotes two subsequent phases, either upswing and consequent downswing, or downswing and consequent upswing.

[9]Gonzalez et al. (2005) use the same algorithm with $\tau_{window} = 6$, $\tau_{cycle} = 15$, and $\tau_{phase} = 5$. Despite the differences, the bull and bear markets in the study by Gonzalez et al. (2005) largely coincide with the bull and bear markets in the study by Pagan and Sossounov (2003).

It is worth noting that the algorithm of Bry and Boschan (1971) exploits the idea that, in order to qualify for a distinct phase, the trend in the stock market price should continue over a substantial period from the previous peak or trough. There is another dating algorithm, proposed by Lunde and Timmermann (2004), which is motivated by the idea that, in order to qualify for a distinct bull or bear phase, the stock market price should change substantially from the previous peak or trough. This dating rule is based on imposing a minimum on the price change since the last peak or trough. Specifically, in this rule one determines a scalar λ_1 that defines the threshold of the movement in stock prices that triggers a switch from a bear state to a bull state, and a scalar λ_2 that defines the threshold for shifts from a bull state to a bear state. When $\lambda_1 = 15\%$ and $\lambda_2 = 10\%$, the algorithm by Lunde and Timmermann (2004) identifies more bull and bear phases than the algorithm by Pagan and Sossounov (2003). However, the bull and bear markets identified by two different rules largely coincide. Our choice of using the dating algorithm by Pagan and Sossounov (2003) is motivated by the following two considerations: this algorithm seems to be more established and recognized in finance literature than the other one, and it does not identify market phases that are relatively short in duration.[10]

Table 9.2 reports the dates of the bull and bear markets over the total sample period from January 1857 to December 2015. In addition, for each market phase the table reports its duration (measured in the number of months) and amplitude (defined as % change in the stock index price from the previous peak or through). Figure 9.2 plots the natural log of the monthly Standard and Poor's Composite stock price index over the two sub-periods: 1857–1943 and 1944–2015. Shaded areas in the figure indicate the bear market phases.

Table 9.2, together with Fig. 9.2, clearly illustrates the major stock market events over the recent 159-year history. The strongest and second-longest bull market in history occurred during the so-called "Roaring Twenties" (August 1923 to August 1929, 295% amplitude, 73-month long), the decade that followed World War I and led to the most severe and third-longest bear market (September 1929 to June 1932, −85% amplitude, 34 month long). The second-largest and the longest bull market was named the "Dot-Com bubble" and happened in the late 1990s (July 1994 to August 2000, 231% amplitude, 74-month long). The second most severe, but relatively short by duration, bear market is known as the "Global Financial Crisis of 2007–2008" (November 2007 to February 2009, −50% amplitude, 16 month long).

---

[10]The algorithm by Lunde and Timmermann (2004) usually produces many market phases with duration of 2–3 months.

Table 9.2 Bull and bear markets over the total sample period 1857–2015

Bull markets			Bear markets		
Dates	Duration	Amplitude	Dates	Duration	Amplitude
			Jan 1857–Oct 1857	10	−45
Nov 1857–Mar 1858	5	45	Apr 1858–Jun 1859	15	−15
Jul 1859–Oct 1860	16	57	Nov 1860–May 1861	7	−24
Jun 1861–Mar 1864	34	176	Apr 1864–Mar 1865	12	−26
Apr 1865–Oct 1866	19	18	Nov 1866–Apr 1867	6	−9
May 1867–Aug 1869	28	33	Sep 1869–Dec 1869	4	−1
Jan 1870–Apr 1872	28	21	May 1872–Nov 1873	19	−22
Dec 1873–Apr 1875	17	2	May 1875–Jun 1877	26	−39
Jul 1877–May 1881	47	119	Jun 1881–Jan 1885	44	−35
Feb 1885–Nov 1886	22	33	Dec 1886–Mar 1888	16	−16
Apr 1888–May 1890	26	18	Jun 1890–Jul 1891	14	−18
Aug 1891–Feb 1892	7	7	Mar 1892–Jul 1893	17	−38
Aug 1893–Aug 1895	25	25	Sep 1895–Aug 1896	12	−27
Sep 1896–Aug 1897	12	35	Sep 1897–Apr 1898	8	−7
May 1898–Apr 1899	12	34	May 1899–Jun 1900	14	−9
Jul 1900–Aug 1902	26	52	Sep 1902–Sep 1903	13	−29
Oct 1903–Jan 1906	28	63	Feb 1906–Oct 1907	21	−36
Nov 1907–Sep 1909	23	57	Oct 1909–Jul 1910	10	−18
Aug 1910–Sep 1912	26	13	Oct 1912–Jul 1914	22	−24
Aug 1914–Oct 1916	27	51	Nov 1916–Nov 1917	13	−31
Dec 1917–Oct 1919	23	29	Nov 1919–Aug 1921	22	−22
Sep 1921–Feb 1923	18	33	Mar 1923–Jul 1923	5	−14
Aug 1923–Aug 1929	73	295	Sep 1929–Jun 1932	34	−85
Jul 1932–Jan 1934	19	83	Feb 1934–Mar 1935	14	−21
Apr 1935–Feb 1937	23	95	Mar 1937–Mar 1938	13	−53
Apr 1938–Dec 1938	9	36	Jan 1939–Apr 1942	40	−38
May 1942–Jun 1943	14	52	Jul 1943–Nov 1943	5	−6
Dec 1943–May 1946	30	64	Jun 1946–Feb 1948	21	−24
Mar 1948–Jun 1948	4	11	Jul 1948–Jun 1949	12	−11
Jul 1949–Dec 1952	42	77	Jan 1953–Aug 1953	8	−12
Sep 1953–Jul 1956	35	112	Aug 1956–Dec 1957	17	−16
Jan 1958–Jul 1959	19	45	Aug 1959–Oct 1960	15	−10
Nov 1960–Dec 1961	14	29	Jan 1962–Jun 1962	6	−20
Jul 1962–Jan 1966	43	60	Feb 1966–Sep 1966	8	−16
Oct 1966–Nov 1968	26	35	Dec 1968–Jun 1970	19	−30
Jul 1970–Apr 1971	10	33	May 1971–Nov 1971	7	−6
Dec 1971–Dec 1972	13	16	Jan 1973–Sep 1974	21	−45
Oct 1974–Dec 1976	27	45	Jan 1977–Feb 1978	14	−15
Mar 1978–Nov 1980	33	58	Dec 1980–Jul 1982	20	−21
Aug 1982–Jun 1983	11	41	Jul 1983–May 1984	11	−7
Jun 1984–Aug 1987	39	115	Sep 1987–Nov 1987	3	−28
Dec 1987–May 1990	30	46	Jun 1990–Oct 1990	5	−15
Nov 1990–Jan 1994	39	49	Feb 1994–Jun 1994	5	−5
Jul 1994–Aug 2000	74	231	Sep 2000–Sep 2002	25	−43
Oct 2002–Oct 2007	61	75	Nov 2007–Feb 2009	16	−50
Mar 2009–Apr 2011	26	71	May 2011–Sep 2011	5	−16
Oct 2011–Dec 2015	51	63			

Notes Duration is measured in the number of months. Amplitudes are defined as % changes in the stock index prices (not adjusted for dividends)

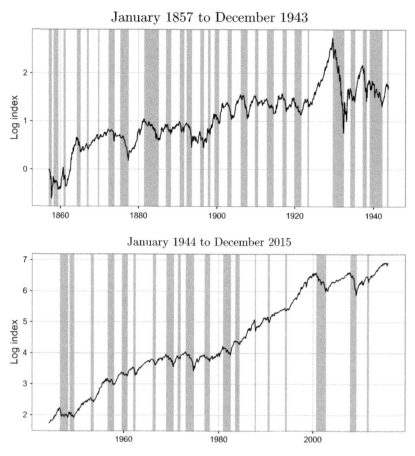

Fig. 9.2 Bull and bear markets over the two historical sub-periods: 1857–1943 and 1944–2015. *Shaded areas* indicate bear market phases

Table 9.3 reports the descriptive statistics of the bull and bear markets for the whole period and the two sub-periods. Over the total period, there were 46 bull markets and 46 bear markets. The first sub-period contains 26 bull markets and 27 bear markets. The second sub-period, which is shorter than the first one, contains 20 bull markets and 19 bear markets. Over the whole period, the average length of a bull market is close to 27 months, whereas the average bear market length is close to 15 months. It is clear that bull markets tend to be longer than bear markets and the durations of phases agree quite closely with those reported by Pagan and Sossounov (2003) and Gonzalez et al. (2005). The average bull market duration exceeds the average bear market duration by a factor of 1.8. The comparison of the lengths of the two stock market phases in the first and the second sub-period suggests that over time bull markets tend to be longer while bear markets tend to be shorter. Specifically,

Table 9.3 Descriptive statistics of bull and bear markets

Statistics	1857–2015		1857–1943		1944–2015	
	Bull	Bear	Bull	Bear	Bull	Bear
Number of phases	46	46	26	27	20	19
Minimum duration	4	3	5	4	4	3
Average duration	26.8	14.7	23.6	16.6	31.3	12.5
Median duration	26	14	23	14	30	12
Maximum duration	74	44	73	44	74	25
Average amplitude, %	59.9	−23.9	56.7	−27.0	63.7	−20.6
Average cum. return, %	89.6	−22.5	88.3	−24.9	90.3	−20.0
Mean monthly return, %	27.6	−21.4	30.5	−21.2	24.4	−21.6
Standard deviation, %	15.3	17.7	17.6	19.3	12.6	14.4

Notes Duration is measured in the number of months. Amplitudes are defined as % changes in the stock index prices (not adjusted for dividends). Cumulative returns, mean monthly return and the standard deviations are computed using the total return (adjusted for dividends)

whereas for the first sub-period the ratio of the average bull market length to the average bear market length amounts to 1.4, for the second sub-period this ratio amounts to 2.5. In other words, this ratio has almost doubled over time. On average, the stock index price increases by 60% during a bull market and decreases by 24% during a bear market. Our results suggest that over time the average amplitude of bull markets tends to increase whereas the average amplitude of bear markets tends to decrease.

All bull markets exhibit positive mean return while all bear markets have negative mean return. Interestingly, over the two sub-periods the mean monthly returns during bear markets were virtually identical. In contrast, the mean monthly return during bull markets was higher in the first sub-period than in the second one. There is economically significant time-variation in the value of standard deviations of both bull and bear market returns across sub-periods of data. Specifically, the market was much more volatile during the first sub-period than during the second one. Somewhat surprisingly,[11] even though in each sub-period the volatility during bear markets was higher than that during bull markets, the difference in volatilities across bull and bear markets is economically insignificant (a similar result is reported by Gonzalez et al. 2005). As an illustration, in the second sub-period the standard deviation of returns was slightly below 13% during bull markets and slightly above 14% during bear markets. This finding implies that bull markets differ from bear markets mainly in terms of mean returns, not in terms of standard deviation of returns.

---

[11] It is customary to assume that a bear market is the low-return high-volatility state of the stock market, whereas a bull market is the high-return low-volatility state of the market.

Our results, together with those obtained previously by Pagan and Sossounov (2003) and Gonzalez et al. (2005), advocate that the properties of cycles in stock prices have significantly changed over time. Yet so far we do not have any scientific evidence of the presence of structural breaks in the parameters and dynamics of bull-bear cycles. Since the presence of structural breaks might be of crucial importance for the ability of a moving average trading strategy to outperform its passive counterpart, we analyze whether there are statistically significant changes in the distribution parameters of bull and bear markets over time. The results of these tests confirm the presence of a structural break in the bull-bear dynamics (see the subsequent appendix for the detailed description of the tests and their results). Specifically, we find statistical evidence that the parameters of the bull and bear markets are different across the two historical sub-periods.

The results of our two structural break analyses (in the growth rate of the index and the dynamics of bull-bear markets) agree with each other and may potentially have important implications for the performance of moving average trading strategies. There is clear scientific evidence that the growth rate of the S&P Composite index has increased in the post-1944 period. In addition, we find evidence that the duration of bull markets has increased over time, whereas the duration of bear markets has decreased over time. Consequently, as compared with the first sub-period, over the second sub-period the index value has been increasing faster and the stock market has been much more often in the Bull state than in the Bear state. Since the superior performance of a moving average trading strategy can appear only as the result of timely identification of Bear market states and undertaking appropriate actions (switching to cash or selling short), it is logical to deduce that we might observe a deterioration in the performance of moving average trading rules (relative to that of the market) over the second sub-period.

9.3   Reducing the Dimensionality of Testing Procedure

Right from the start, we reduced the number of tested rules to 4 (MOM, MAC, MAE, and MACD). We do not need more rules to generate the most common shapes of the price-change weighting function that is used to compute the trading signal in a moving average rule.[12] However, a practical realization of 3 out of 4 rules requires choosing a particular moving average. Even though

---

[12]In fact, we need only the MOM, MAC, and MACD rules. We add the MAE rule in order to see whether it outperforms the MAC rule; see the discussion in the beginning of the previous part of the book.

we decided to employ only ordinary moving averages, the number of possible combinations (of a trading rule and a moving average) becomes relatively large. In addition, when a trading rule generates a Sell signal, there are two possible actions: either move to cash or sell short the stocks. Finally, because there are several alternative performance measures, the selection of the best trading strategy may depend on the choice of performance measure. To reduce the dimensionality of testing procedure, in this section we answer the following questions: Does the choice of performance measure influence the selection of the best trading strategy? Does the choice of moving average influence the performance of the best trading strategy? Is it sensible to consider the strategy with short sales?

9.3.1   Does the Choice of Performance Measure Influence the Selection of Trading Strategy?

The most widely known performance measure is the Sharpe ratio, which is a reward-to-risk ratio where the risk is measured by the standard deviation of returns. The choice of the Sharpe ratio as a performance measure is fully justified when returns are normally distributed.[13] However, empirical literature frequently documents that financial asset return distributions deviate from normality. The results of our tests also suggest that we can reject the hypothesis that the returns on the S&P Composite index are normally distributed (see Sect. 9.1). When return distributions are non-normal, it is commonly believed that the performance cannot be adequately evaluated using the Sharpe ratio. As a result of this belief, researchers have proposed a vast number of different performance measures that try to take into account the non-normality of return distributions (see, for example, Cogneau and Hübner 2009, for a good review of different performance measures).

Specifically, the Sharpe ratio is often criticized on the grounds that the standard deviation appears to be an inadequate measure of risk. In particular, the standard deviation similarly penalizes both the downside risk and the upside return potential. Because a moving average trading strategy is supposed to provide downside protection and upside participation, it is natural to think that the Sharpe ratio is inappropriate for the performance measurement of these strategies. The Sortino ratio (see Sortino and Price, 1994), which uses downside deviation as a risk measure, seems to be a much more reasonable performance measure than the Sharpe ratio.

---

[13]Yet, recall that the justification of the Sharpe ratio is based on a number of additional assumptions. Besides the normality of return distributions, one has to assume the existence of a risk-free asset and the absence of any limitations on borrowing and lending.

However, there is another stream of research that advocates that the choice of performance measure does not influence the evaluation of risky portfolios. For example, Eling and Schuhmacher (2007), Eling (2008), and Auer (2015) computed the rank correlations between the rankings produced by a set of alternative performance measures (including the Sharpe ratio), and found that the rankings are extremely positively correlated. To check whether the choice of performance measure influences the selection of the best trading strategy, we conduct an empirical study to shed light on the issue of performance measure choice in the context of moving average trading strategies.

Our empirical study is conducted in the following manner. First, we select a simple trading rule (MOM(n), P-SMA(n), P-LMA(n), or P-EMA(n)) whose performance depends on the choice of a single parameter, n, the size of the window used to compute the trading signal. Then we simulate the returns to this trading rule over the total historical sample and different values of n beginning from $n_{min} = 2$ and ending with $n_{max} = 25$ (and accounting for 0.25% one-way transaction costs). That is, we simulate the returns to 24 different trading strategies. Next, we select a performance measure (Mean excess return, Sharpe ratio, or Sortino ratio), evaluate the performance of each trading strategy (over the period from January 1860 to December 2015), and rank each trading strategy. Specifically, the best performing strategy is assigned rank 1, the next best performing strategy is assigned rank 2, and so on down to 24. The outcome of this procedure is three sets of ranks; ranks according to the Mean excess return performance criterion, ranks according to the Sharpe ratio criterion, and ranks according to the Sortino ratio criterion. Finally, we calculate rank correlations between these sets of ranks. Following Eling and Schuhmacher (2007), Eling (2008), and Auer (2015) we use the Spearman rank correlation coefficients ρ as a nonparametric measure of rank correlation. As for any correlation coefficient, the value of ρ is restricted to lie within two boundaries $-1 \leq \rho \leq 1$. If two sets of ranks are identical, then $\rho = 1$. If two sets of ranks are completely different, then $\rho = -1$.

For each trading rule in this study, Table 9.4 reports the Spearman rank correlation coefficients between three alternative performance measures. Apparently, all performance measures display a very high rank correlation with respect to each other. The rank correlation coefficient varies between 0.97 and 1.00. For the sake of illustration of ranking of different trading strategies, Table 9.5 lists the top 10 strategies according to each performance measure. It is worth

Table 9.4 Rank correlations based on different performance measures

	MOM			P-SMA		
	Excret	Sharpe	Sortino	Excret	Sharpe	Sortino
Excret	1.00			1.00		
Sharpe	0.99	1.00		0.99	1.00	
Sortino	0.99	0.99	1.00	0.97	0.99	1.00
	P-LMA			P-EMA		
	Excret	Sharpe	Sortino	Excret	Sharpe	Sortino
Excret	1.00			1.00		
Sharpe	1.00	1.00		1.00	1.00	
Sortino	1.00	1.00	1.00	0.98	0.98	1.00

Notes For each trading rule (MOM(n), P-SMA(n), P-LMA(n), and P-EMA(n)), this table reports the Spearman rank correlation coefficients between three alternative performance measures: Mean excess returns (Excret), Sharpe ratio (Sharpe), and Sortino ratio (Sortino). The values are rounded to the second decimal place. Because of the symmetry of the correlation matrix, we do not report its upper-right triangle

noting that the trading strategies in this list are the best trading strategies in a back test. Observe, for example, that the P-SMA(12) strategy appears to be the best trading strategy (among all tested P-SMA(n) strategies) in a back test regardless of the measure used to evaluate the performance. The most popular among practitioners P-SMA(10) strategy is also among the top 10 best strategies, but its rank depends on the choice of performance measure. Also observe that according to the results reported in Table 9.4, all rank correlations amount to 1.00 for the P-LMA(n) rule. However, the list of the top 10 P-LMA(n) strategies in Table 9.5 suggests that the ranks are not fully identical. The explanation for this seemingly conflicting information reported in these two tables is that we round the values of rank correlations to the second decimal place. In reality, when a rank correlation between two performance measures is reported to be 1.00, its value lies in between 0.995 and 1.000.

The main conclusion that we can draw from this empirical study is that the choice of performance measure does not affect the ranking of moving average trading strategies as much as one would expect after studying the performance measurement literature. Our findings are in complete agreement with the findings reported previously by Eling and Schuhmacher (2007), Eling (2008), and Auer (2015). The implications of our results are as follows. From a practical point of view, the choice of performance measure does not have a crucial influence on the relative evaluation of moving average trading strategies. Taking into account that the Sharpe ratio is the best known and best understood

Table 9.5 The top 10 strategies according to each performance measure

	MOM			P-SMA		
Rank	Excret	Sharpe	Sortino	Excret	Sharpe	Sortino
1	MOM(11)	MOM(6)	MOM(6)	P-SMA(12)	P-SMA(12)	P-SMA(12)
2	MOM(6)	MOM(11)	MOM(11)	P-SMA(15)	P-SMA(15)	P-SMA(15)
3	MOM(8)	MOM(8)	MOM(8)	P-SMA(17)	P-SMA(16)	P-SMA(16)
4	MOM(10)	MOM(9)	MOM(10)	P-SMA(16)	P-SMA(17)	P-SMA(10)
5	MOM(12)	MOM(10)	MOM(9)	P-SMA(14)	P-SMA(14)	P-SMA(17)
6	MOM(9)	MOM(13)	MOM(12)	P-SMA(13)	P-SMA(11)	P-SMA(11)
7	MOM(13)	MOM(12)	MOM(5)	P-SMA(11)	P-SMA(10)	P-SMA(14)
8	MOM(5)	MOM(5)	MOM(7)	P-SMA(18)	P-SMA(13)	P-SMA(13)
9	MOM(7)	MOM(7)	MOM(13)	P-SMA(10)	P-SMA(18)	P-SMA(18)
10	MOM(14)	MOM(14)	MOM(16)	P-SMA(19)	P-SMA(9)	P-SMA(9)
	P-LMA			P-EMA		
Rank	Excret	Sharpe	Sortino	Excret	Sharpe	Sortino
1	P-LMA(22)	P-LMA(22)	P-LMA(22)	P-EMA(13)	P-EMA(13)	P-EMA(13)
2	P-LMA(20)	P-LMA(20)	P-LMA(20)	P-EMA(14)	P-EMA(14)	P-EMA(11)
3	P-LMA(24)	P-LMA(21)	P-LMA(21)	P-EMA(12)	P-EMA(11)	P-EMA(14)
4	P-LMA(21)	P-LMA(24)	P-LMA(24)	P-EMA(11)	P-EMA(12)	P-EMA(12)
5	P-LMA(23)	P-LMA(23)	P-LMA(23)	P-EMA(15)	P-EMA(10)	P-EMA(10)
6	P-LMA(25)	P-LMA(18)	P-LMA(18)	P-EMA(10)	P-EMA(15)	P-EMA(15)
7	P-LMA(18)	P-LMA(25)	P-LMA(25)	P-EMA(9)	P-EMA(8)	P-EMA(8)
8	P-LMA(17)	P-LMA(19)	P-LMA(19)	P-EMA(8)	P-EMA(9)	P-EMA(9)
9	P-LMA(19)	P-LMA(17)	P-LMA(17)	P-EMA(16)	P-EMA(16)	P-EMA(7)
10	P-LMA(16)	P-LMA(16)	P-LMA(16)	P-EMA(17)	P-EMA(7)	P-EMA(16)

Notes For each trading rule (MOM(n), P-SMA(n), P-LMA(n), and P-EMA(n)), this table reports the top 10 strategies according to three alternative performance measures: Mean excess returns (Excret), Sharpe ratio (Sharpe), and Sortino ratio (Sortino). The performance of all strategies is evaluated over the period from January 1860 to December 2015

performance measure, it might be considered superior to other performance measures from a practitioner's point of view. We thus conclude that from a practical point of view the choice of performance measure does not influence the ranking of trading strategies. In the subsequent analysis, we will employ only the Sharpe ratio for measuring the performance of trading strategies.

9.3.2 To Short or Not to Short?

When a new monthly closing price becomes available, a moving average trading rule generates the trading signal (Buy or Sell) for the subsequent month. A Buy signal is always a signal to invest in the stocks or stay invested in the stocks. When a Sell signal is generated after a Buy signal, there are two alternative strategies: either (1) sell the stocks and invest the proceeds in cash or (2) sell the stocks, additionally sell short the stocks, and invests all proceeds (from the sale and short sale) in cash. If a moving average trading rule correctly predicts bear markets, the first strategy just protects the trader from losses, whereas the second strategy allows the trader to profit from a drop in stock prices. However, if the precision of identification of bear markets is low, the trader often loses money on short sales. Therefore the first strategy possesses an advantage over the second strategy when market timing is poor.

To find out whether it is sensible to consider the strategy with short sales in in-sample and out-of-sample tests, the following empirical study is conducted. First, we simulate the returns to two simple trading rules, MOM(n) and P-SMA(n), with and without short sales. The returns to different trading strategies are simulated over the total historical sample and different values of n beginning from $n_{min} = 2$ and ending with $n_{max} = 25$ (accounting for 0.25% one-way transaction costs). Next, using the Sharpe ratio as performance measure, we evaluate to which extent each trading strategy outperforms the passive strategy. Formally, for each strategy we compute $\Delta = SR_{MA} - SR_{BH}$ where SR_{MA} and SR_{BH} are the Sharpe ratios of the moving average strategy and the buy-and-hold strategy respectively. Finally, we rank each trading strategy according to its outperformance, from best to worst.

Table 9.6 reports the top 10 strategies for each trading rule. Specifically, the left panel in the table reports the performance of the top 10 trading strategies without the short sales, whereas the right panel reports the performance of the top 10 trading strategies with short sales. Obviously, the results reported in this table clearly suggest that short sales significantly deteriorate the performance of trading rules. In particular, whereas all top 10 trading strategies without short sales outperform the buy-and-hold strategy, all top 10 trading strategies with short sales underperform the buy-and-hold strategy. It is worth mentioning that the performance of all strategies in this empirical study was evaluated using the in-sample methodology. Consequently, with short sales even the best trading rule in a back test fails to outperform the passive benchmark. Since the in-sample performance of any trading rule overestimates its real-life performance, we can confidently say that the out-of-sample performance of trading rules with short sales is inferior to the performance of the buy-and-hold strategy.

In the end of this section we would like to elaborate a bit more on the practical implications of our results. In order a trading strategy with short sales to outperform its counterpart without short sales, near-perfect market timing is required. Since our results show that the performance of strategies with short sales is much poorer than that of the strategies without short sales, this finding suggests that even in a back test the best trading strategy identifies the bull and bear market phases with poor accuracy. Why this accuracy is poor? The answer

Table 9.6 Comparative performance of trading strategies with and without short sales

| Rank | Switch to Cash | | Sell Short | |
	Strategy	Δ	Strategy	Δ
	Momentum rule			
1	MOM(6)	0.15	MOM(11)	−0.04
2	MOM(11)	0.14	MOM(6)	−0.06
3	MOM(8)	0.13	MOM(8)	−0.10
4	MOM(9)	0.09	MOM(10)	−0.10
5	MOM(10)	0.09	MOM(12)	−0.12
6	MOM(13)	0.07	MOM(9)	−0.14
7	MOM(12)	0.07	MOM(13)	−0.14
8	MOM(5)	0.07	MOM(5)	−0.17
9	MOM(7)	0.07	MOM(7)	−0.17
10	MOM(14)	0.04	MOM(14)	−0.19
	Price-SMA rule			
1	P-SMA(12)	0.15	P-SMA(12)	−0.06
2	P-SMA(15)	0.14	P-SMA(15)	−0.07
3	P-SMA(16)	0.14	P-SMA(17)	−0.07
4	P-SMA(17)	0.13	P-SMA(16)	−0.07
5	P-SMA(14)	0.13	P-SMA(14)	−0.08
6	P-SMA(11)	0.13	P-SMA(13)	−0.08
7	P-SMA(10)	0.13	P-SMA(11)	−0.09
8	P-SMA(13)	0.13	P-SMA(18)	−0.09
9	P-SMA(18)	0.12	P-SMA(10)	−0.10
10	P-SMA(9)	0.10	P-SMA(19)	−0.12

Notes The performance of all strategies is evaluated over the period from January 1860 to December 2015. $\Delta = SR_{MA} - SR_{BH}$ where SR_{MA} and SR_{BH} are the Sharpe ratios of the moving average strategy and the buy-and-hold strategy respectively. The Sharpe ratio of the buy-and-hold strategy amounts to 0.39

can be provided by examining the timing properties of the best trading strategy in a back test. Let us consider the best trading strategy in the P-SMA(n) rule. In the best trading strategy the size of the averaging window amounts to 12 months. That is, in this strategy the trend in the stock price is detected using the 12-month simple moving average. We know from Chap. 3 that the average lag time of this moving average equals $(n - 1)/2$. Consequently, in the best trading strategy a turning point in the stock price trend is identified with the average delay of 5.5 months. Therefore if a bear market lasts 5–6 months, then roughly the best trading strategy generates a wrong Buy signal during the whole bear market and afterwards generates a wrong Sell signal during the first 5–6 months of the subsequent bull market. Indeed, the duration of a stock price trend should be long enough to make the trend following strategy profitable. Since over the total sample period the median duration of a bear market equals 14 months, there is a good reason to think (as a ballpark estimate) that even the best trading strategy in a back test underperforms the buy-and-hold strategy during half of all bear markets (that is, during bear markets with duration shorter than 14 months).

Overall, our conclusion is that the short selling strategy is risky and does not pay off. In addition, there are some other practical complications with implementation of this strategy. First, this strategy involves significant expenses. In our empirical study we accounted for the double transaction costs only. In reality, there are additional short borrowing costs. We assumed that the trader can always sell short stocks, but in reality regulators may impose bans on short sales to avoid panic and unwarranted selling pressure. We supposed that stocks can be sold short as long as the trader wants. In real markets, because short selling means selling borrowed stocks, the trader can be forced to cover the short sale if the lender wants the stocks back. Therefore this strategy is not only highly risky, but very expensive and sometimes impossible to implement.

9.3.3   Does the Choice of Moving Average Influence the Performance of Trading Strategy?

The three ordinary moving averages are SMA(n), LMA(n), and EMA(n). In the first part of the book we established that both SMA(n) and EMA(n) have the same tradeoff between the average lag time and smoothness, whereas LMA(n) has a slightly better tradeoff between the average lag time and smoothness. In addition, our experiments revealed that, at least in the case where the trend has no noise, for the same average lag time the EMA(n) has the shortest delay time in turning point identification, whereas SMA(n) has the longest delay time in turning point identification. Therefore one naturally expects that the performance of trading rules with either EMA(n) or LMA(n) should be superior as compared to that of trading rules with SMA(n). However, our numerous graphical illustrations provided in the first part of the book demonstrated that: (1) all ordinary moving averages move close together when they have the same average lag time; (2) the price-change weighting functions of all ordinary moving averages differ only a little.

Our goal in this section is to evaluate the comparative performance of trading rules with different types of ordinary moving averages and, for each specific rule, find out whether there is a moving average that is clearly superior to the others. In other words, we want to answer the question of whether the choice of moving average influences the performance of trading rules. This is done in the following manner. First, we select a trading rule and, using the total sample of data, simulate the returns to this rule with three different types of moving averages. For example, we select the P-MA(n) rule and simulate the returns to this rule (accounting for 0.25% one-way transaction costs) with

SMA, LMA, and EMA. For each type of moving average, we vary the value of n in $[2, 25]$. Thus, for each trading rule and each moving average we simulate the returns to 24 trading strategies. Next, using the Sharpe ratio as performance measure, we evaluate to which extent each trading strategy (out of totally $24 \times 3 = 72$ strategies) outperforms the passive strategy. Formally, for each strategy we compute $\Delta = SR_{MA} - SR_{BH}$ where SR_{MA} and SR_{BH} are the Sharpe ratios of the moving average strategy and the buy-and-hold strategy respectively. Finally, we rank each trading strategy according to its outperformance, from best to worst. Besides the P-MA rule, we use the MAC rule, the MAE rule, and the MACD rule.

Table 9.7 reports the top 10 best strategies for each trading rule. Rather surprisingly, contrary to the common belief that LMA and EMA are superior to SMA, for 3 out of 4 trading rules a trading strategy with SMA provides either the best performance or one of the best performances. Specifically, for the P-MA rule the three best performing strategies are based on using SMA. Similarly, for the MAC rule the two best performing strategies are based on using SMA. For the MAE rule the strategy with SMA is ranked 3rd, but its performance is virtually the same as that of the two top strategies that employ LMA. For the P-MA, MAC, and MAE rules the strategies with EMA are virtually absent from the top 10 best performing strategies (yet they appear more frequently in the top 20 best performing strategies). In contrast, for the MACD rule all top 10 strategies are based on using EMA.

One should keep in mind, however, that the results of our study, as the results of any empirical study, are dataset-specific and data frequency-specific. However, when the long-term monthly data on the S&P Composite index are used, the results seem to be clear-cut. In particular, our results advocate that the choice of moving average is of little importance. When either the P-MA, MAC, or MAE rule is used, trading strategies with either SMA or LMA perform virtually similar. Even though the performance of strategies with EMA is worse than that with SMA and LMA, the difference in performances is rather small. For example, the best P-SMA strategy has a Sharpe ratio of 0.54 (0.39 Sharpe ratio of the buy-and-hold strategy plus 0.15 outperformance), whereas the best P-EMA strategy has a Sharpe ratio of 0.52. Even for the MACD rule the situation is exactly the same: the best EMACD strategy has a Sharpe ratio of 0.55, whereas the best SMACD strategy has a Sharpe ratio of 0.52.

The results of this empirical study reveal that the choice of moving average does not affect the performance of moving average trading strategies as much as one would expect by examining the price weighting functions of different moving averages. Many traders believe that LMA and EMA possess better properties than SMA. In reality, it turns out that "better" is only in the eye of

Table 9.7 Comparative performance of trading rules with different types of moving averages

Rank	Strategy	Δ	Strategy	Δ
	P-MA rule		**MAC rule**	
1	P-SMA(12)	0.15	SMAC(2,10)	0.16
2	P-SMA(15)	0.14	P-SMA(12)	0.15
3	P-SMA(16)	0.14	LMAC(2,20)	0.14
4	P-LMA(22)	0.13	LMAC(2,21)	0.14
5	P-SMA(17)	0.13	P-SMA(15)	0.14
6	P-LMA(20)	0.13	SMAC(2,12)	0.14
7	P-SMA(14)	0.13	EMAC(4,8)	0.14
8	P-SMA(11)	0.13	P-SMA(16)	0.14
9	P-SMA(10)	0.13	P-LMA(22)	0.13
10	P-SMA(13)	0.13	LMAC(2,18)	0.13
	MAE rule		**MACD rule**	
1	LMAE(21,0.25)	0.15	EMACD(8,23,10)	0.16
2	LMAE(21,0.5)	0.15	EMACD(10,23,8)	0.16
3	P-SMA(12)	0.15	EMACD(7,22,10)	0.16
4	SMAE(15,0.75)	0.15	EMACD(10,22,7)	0.16
5	SMAE(12,0.25)	0.15	EMACD(8,24,8)	0.16
6	SMAE(16,0.5)	0.15	EMACD(9,23,9)	0.16
7	SMAE(12,1)	0.14	EMACD(6,21,12)	0.15
8	LMAE(14,2.5)	0.14	EMACD(12,21,6)	0.15
9	SMAE(12,1.25)	0.14	EMACD(8,23,8)	0.15
10	LMAE(16,1.5)	0.14	EMACD(9,19,9)	0.15

Notes The performance of all strategies is evaluated over the period from January 1860 to December 2015. $\Delta = SR_{MA} - SR_{BH}$ where SR_{MA} and SR_{BH} are the Sharpe ratios of the moving average strategy and the buy-and-hold strategy respectively. The Sharpe ratio of the buy-and-hold strategy amounts to 0.39

the beholder. The better properties of LMA and EMA, as compared to those of SMA, do not show up in empirical tests. Our main conclusion from this empirical study is that the choice of moving average is irrelevant. That is, from a practical point of view, the choice of moving average does not have a crucial influence on the performance of moving average trading strategies. Taking into account that the SMA is the simplest, best known, and best understood moving average, it might be considered superior to other moving averages from a practitioner's point of view. Motivated by this conclusion, in the subsequent tests we will employ only SMA in the P-MA, MAC, and MAE rules. However, we will implement the MACD rule with EMA. The reason for the latter choice is that the MACD rule traditionally uses EMA.

9.3.4   A Brief Summary of Results

In this section we performed three empirical studies. The results of these studies allow us to make the following conclusions:

- The short selling strategy is risky and does not pay off. Specifically, the performance of the short selling strategy is substantially worse than the performance of the corresponding strategy where the trader switches to cash (or stays invested in cash) after a Sell signal is generated.
- From a practical point of view, the choice of performance measure does not influence the performance ranking of trading strategies. Therefore the Sharpe ratio, which has become the industry standard for measuring risk-adjusted performance, is superior to other performance measures from a practitioner's point of view.
- From a practical point of view, the choice of moving average does not have a crucial influence on the performance of moving average trading strategies. In particular, regardless of the choice of moving average, the performance of the best trading strategy in a back test remains virtually intact. In this regard, the Simple Moving Average can be preferred as the simplest, best known and best understood moving average.

9.4   Back-Testing Trading Rules

The results, presented in the previous section, give us some information about the best performing strategies in a back test over the total historical sample of data. In particular, among the set of P-SMA, SMAC, and SMAE rules, the best performing strategies over the total sample are the P-SMA(12) and SMAC(2,10) strategies. The goal of this section is to perform a deeper analysis of the best performing moving average strategies in a back test.

The following set of rules are tested:

$MOM(n)$ for $n \in [2, 25]$, totally 24 trading strategies;

$SMAC(s, l)$ for $s \in [1, 12]$ and $l \in [2, 25]$, totally 222 trading strategies;

$SMAE(n, p)$ for $n \in [2, 25]$ and $p \in [0.25, 0.5, \ldots, 5.0]$, totally 480 trading strategies;

$EMACD(s, l, n)$ for $s \in [1, 12], l \in [2, 25]$, and $n \in [2, 12]$, totally 2,442 trading strategies.

Table 9.8 Top 10 best trading strategies in a back test

Rank	Strategy	Δ	Strategy	Δ	Strategy	Δ
	1860–2015		1860–1943		1944–2015	
1	EMACD(8,23,10)	0.16	EMACD(9,23,9)	0.20	SMAC(2,10)	0.15
2	EMACD(10,23,8)	0.16	EMACD(8,23,10)	0.19	SMAC(2,12)	0.15
3	EMACD(7,22,10)	0.16	EMACD(10,23,8)	0.19	SMAC(2,11)	0.15
4	EMACD(10,22,7)	0.16	EMACD(8,21,11)	0.19	SMAE(8,1.25)	0.14
5	EMACD(8,24,8)	0.16	EMACD(11,21,8)	0.19	SMAE(8,1.5)	0.13
6	EMACD(9,23,9)	0.16	EMACD(8,24,10)	0.19	SMAE(11,0.75)	0.13
7	SMAC(2,10)	0.16	EMACD(10,24,8)	0.19	EMACD(9,18,10)	0.12
8	EMACD(6,21,12)	0.15	EMACD(10,15,10)	0.19	EMACD(10,18,9)	0.12
9	EMACD(12,21,6)	0.15	EMACD(9,17,12)	0.19	SMAE(11,0.5)	0.12
10	EMACD(8,23,8)	0.15	EMACD(12,17,9)	0.19	P-SMA(12)	0.12

Notes $\Delta = SR_{MA} - SR_{BH}$ where SR_{MA} and SR_{BH} are the Sharpe ratios of the moving average strategy and the buy-and-hold strategy respectively

The overall number of tested trading strategies amounts to 3,168. The returns to all strategies are simulated accounting for 0.25% one-way transaction costs. In all strategies a Sell signal is a signal to leave the stocks and move to cash (or stay invested in cash). The performance of all strategies is measured using the Sharpe ratio.

Table 9.8 reports the top 10 best trading strategies in a back test over the total sample, as well as over the first and the second part of the sample. Apparently, over the first part of the historical sample, from 1860 to 1943, the trading strategies based on the EMACD rule show the best performance in a back test. As a matter of fact, it should be of no surprise that the EMACD rule is over-represented among the top 10 best performing rules; this rule is very flexible and easier to fit to data than the other rules because the shape of its price-change weighting function is determined by 3 parameters (as a result, the number of tested EMACD strategies is much greater than the number of all other tested strategies). Over the second part of the sample, from 1944 to 2015, the SMAC(2,10) strategy shows the best performance in a back test. This rule is also among the top 10 best performing strategies over the total historical sample from 1860 to 2015.

Figure 9.3 shows the shapes of the price-change weighting functions of the best trading strategies in a back test. Specifically, it plots the price-change weighting function of the EMACD(8,23,10) trading strategy which performs best over the total historical sample, as well as the price-change weighting function of the SMAC(2,10) trading strategy which is among the top 10 best trading strategies over the total sample. All other trading strategies, that are among the top 10 trading strategies over the total sample, belong to the

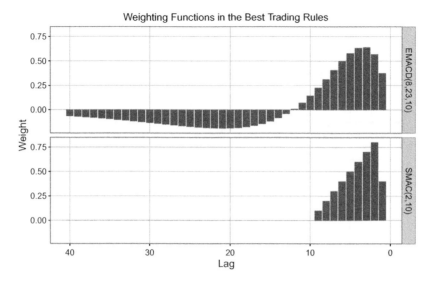

Fig. 9.3 The shapes of the price-change weighting functions of the best trading strategies in a back test

EMACD rule; the shapes of their price-change weighting functions are similar to that of the EMACD(8,23,10) trading strategy.

The price-change weighting function of the SMAC(2,10) strategy has a hump-shaped form, whereas the price-change weighting function of the EMACD(8,23,10) strategy has a damped waveform. Nevertheless, a visual observation reveals that the price-change weighting functions of both EMACD(8,23,10) and SMAC(2,10) strategies look quite similar for the first 9 lags. While the SMAC rule is a genuine trending rule, the EMACD rule performs best when prices are mean-reverting. Figure 9.2, upper panel, helps explain why the EMACD rule performed very well over the first part of the sample. In particular, as shown in the upper panel of this figure, the upswings and downswings in the S&P Composite index appear to have followed each other with sufficient regularity over the first part of the sample. Over the second part of the sample, on the other hand, this regularity disappeared. As a result of this disappearance, the EMACD rule lost its advantage over the SMAC rule.

Over the total historical sample, the performances of the EMACD(8,23,10) strategy and the SMAC(2,10) strategy differ marginally. Both trading strategies outperform the buy-and-hold strategy[14]; the difference in the Sharpe ratio of the best moving average trading strategy and the Sharpe ratio of the buy-and-hold strategy amounts to 0.15–0.16. However, a very prominent feature of

---

[14] It is worth repeating, however, that the performance of the best trading rule in a back test overestimates the real-life performance, because it is upward biased.

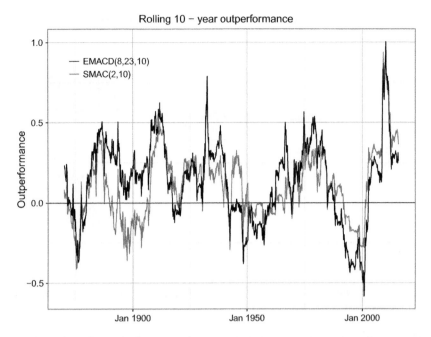

Fig. 9.4 Rolling 10-year outperformance produced by the best trading strategies in a back test over the total historical period from January 1860 to December 2015. The first point in the graph gives the outperformance over the first 10-year period from January 1860 to December 1869. Outperformance is measured by $\Delta = SR_{MA} - SR_{BH}$ where SR_{MA} and SR_{BH} are the Sharpe ratios of the moving average strategy and the buy-and-hold strategy respectively

the outperformance generated by a moving average trading strategy is the fact that this outperformance is very uneven over time. This distinctive feature of the outperformance was, for the first time, emphasized in the paper by Zakamulin (2014). Therefore, as argued in Zakamulin (2014), the traditional performance measurement, that uses a single number for outperformance,[15] is very misleading. This is because a single number for outperformance creates a wrong impression that outperformance is time-invariant, whereas in reality it varies dramatically over time.

Figure 9.4 plots the rolling 10-year outperformance produced by the best trading strategies in a back test over the total historical period from January 1860 to December 2015. Specifically, the first point in the graph gives the outperformance over the first 10-year period from January 1860 to December 1869; the second point gives the outperformance over the second 10-year period from February 1860 to January 1870, etc. Apparently, the conclusions

---

[15] Furthermore, such a performance is usually measured over a very long-term horizon which is beyond the investment horizon of most individual investors.

that can be drawn from this plot are clear-cut: the outperformance varies dramatically over time; there are long periods where even the best trading strategies in a back test underperform the buy-and-hold strategy. For example, the SMAC(2,10) trading strategy, which performed best over the second part of the sample, underperformed the buy-and-hold strategy over approximately 20-year long period from 1982 to 2001.

It is worth noting that the above results on the best performing trading strategies answer the following question: which trading strategy delivers the best performance if the trader sticks to one single trading strategy over the whole tested period? The other interesting question, which is not answered by the back tests performed above, is whether the optimal trading strategy is time-invariant. In other words, it is interesting to find out whether over any given historical period the same trading strategy delivers the best performance. In order to find this out, we perform the following "rolling" back test. In particular, we use a 10-year rolling window and, for each overlapping period of 10 years (over the total sample from 1860 to 2015), find the best trading strategy in a back test. After finding the best trading strategies in all 10-year windows,[16] we count the frequency of each trading strategy. That is, we count over how many rolling windows a specific trading strategy is the best performing strategy.

For the sake of simplicity and clarity, the set of tested trading rules consists of only the MOM(n) rule, the SMAC(s, l) rule, and the buy-and-hold rule denoted by B&H(). We add the buy-and-hold strategy because the previous test reveals that the trend following strategies do not always outperform the buy-and-hold strategy. We do not employ the EMACD(s, l, n) rule because this rule is way too flexible and, as a result, the number of possible trading strategies in this rule exceeds by far the number of possible trading strategies in all other rules.

Figure 9.5 plots the top 20 most frequent trading strategies in a rolling back test. Apparently, the SMAC(2,10) strategy is not among the 20 most frequent trading strategies. The first most frequent trading strategy is the MOM(5) strategy, the second most frequent trading strategy is the MOM(11) strategy. Interestingly, the buy-and-hold strategy is the third most frequent trading strategy in a rolling back test. The two conclusions that can be drawn from these results are as follows. First, there is no single strategy that delivers the best performance over any arbitrarily chosen historical period. Second, over short- to medium-term horizons quite often a moving average trading strategy cannot beat the buy-and-hold strategy even in a back test.

The strategies that are among the 20 most frequent trading strategies are not completely unrelated to each other. In fact, there are many strategies that differ

---

[16]The total number of overlapping 10-year windows amounts to 1752.

20 most frequent trading rules

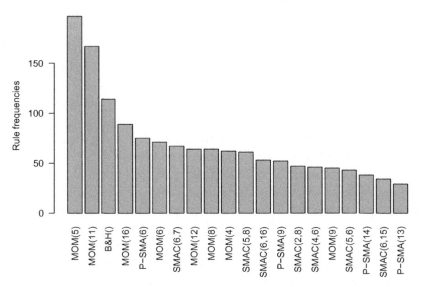

Fig. 9.5 The top 20 most frequent trading rules in a rolling back test. A 10-year rolling window is used to select the best performing strategies over the full sample period from 1860 to 2015

only a little. Examples are: MOM(11) and MOM(12) strategies, SMAC(6,15) and SMAC(6,16) strategies, MOM(4) and MOM(5) strategies, etc. In order to analyze the relationship between the most frequent trading strategies, we compute the correlation coefficients between the returns to these strategies and, on the basis of the correlation matrix, we construct a cluster dendrogram. This dendrogram is depicted in Fig. 9.6. A dendrogram is a visual representation of the correlation matrix. The individual components, in our context they are the 20 most frequent trading strategies, are arranged along the bottom of the dendrogram and referred to as "leaf nodes". Individual components are joined into clusters with the join point referred to as a "node".

The vertical axis in a dendrogram is labelled "distance" and refers to a distance measure between individual components or "clusters". The height of the node can be thought of as the distance value between the right and left sub-branch clusters. The distance measure between two clusters is calculated as one minus the correlation coefficient times 100 (that is, $D = (1 - C) \times 100$, where D and C denote the distance and the correlation coefficient respectively). The smaller the distance, the higher the correlation coefficient. For example, the dendrogram reveals that the returns to the P-SMA(13) and P-SMA(14) strategies are highly correlated. This result comes as no surprise because both

Cluster dendrogram for 20 most frequent trading rules

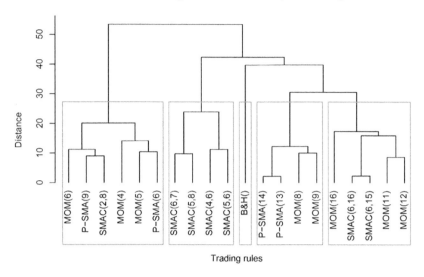

Fig. 9.6 Cluster dendrogram that shows the relationship between the 20 most frequent trading strategies in a rolling back test

strategies belong to the same P-SMA strategy and the sizes of the averaging windows in these strategies differ by one monthly observation.

We separate all trading strategies into a few distinct clusters and draw rectangles around the branches of a dendrogram highlighting the corresponding clusters. Whereas the buy-and-hold strategy represents an individual cluster in this cluster dendrogram, all moving average trading strategies can be divided between 4 clusters. 3 out of 4 of these clusters are comprised of typical moving average trading strategies for which the price-change weighting function has either equally-weighted, decreasing, or hump-shaped form. The main difference between these clusters is in the size of the averaging window (or average lag time). Specifically, these clusters are comprised of: (1) strategies with short averaging window (examples are MOM(4), MOM(5), and MOM(6)), (2) strategies with medium averaging window (examples are MOM(8) and MOM(9)), and (3) strategies with long averaging window (examples are MOM(11) and MOM(12)). Surprisingly, the 4th cluster is comprised of strategies for which the price-change weighting function has increasing form.[17] That is, this type of a price-change weighting function assigns larger weights to more distant price changes (example is the SMAC(6,7) strategy).

---

[17] Recall a discussion in Sect. 5.5. Specifically, in the SMAC(s, l) rule, when s is close to l, the hump (or the top) of the price-change weighting function is located closer to the most distant price change. When $s = l - 1$, the shape of the price-change weighting function has a distinct increasing form.

In principle, the shorter the sample the larger the effect of randomness and, consequently, the larger the data-mining bias. However, we believe that a big diversity of the set of the best trading rules in a rolling back test cannot be attributed to randomness alone. The results of the rolling back test suggest that there is no single rule that performs best in any given period. The type of the optimal trading rule is changing over time. Sometimes trading rules with a short average lag time perform best, other times trading rules with a long average lag time perform best. Therefore the optimality of the SMAC(2,10) strategy over the very long historical sample period appears likely due to the fact that this strategy is "optimal on average" over all possible sub-periods.

9.5   Forward-Testing Trading Rules

9.5.1   Forward or Walk-Forward?

In an out-of-sample testing procedure, in-sample segment of data can be either rolling or expanding. The use of a rolling in-sample window in out-of-sample tests (this technique is called walk-forward testing) is justified when the market's dynamics is changing over time. The results reported in the previous section advocate that, over short- to medium-term horizons, the best trading rule in a back test is changing over time. Therefore, implementing forward tests of moving average trading rules with a rolling in-sample window can potentially produce better out-of-sample performance of trading rules. The goal of this section is to test whether the out-of-sample performance of moving average trading rules depends on forward-testing technique (use of either expanding or rolling in-sample window).

The set of tested trading rules is the same as that described in Sect. 9.4. The overall number of tested trading strategies amounts to 3,168. Recall that in a forward test a trading signal at month-end equals the trading signal of the strategy (1 out of overall 3,168 available strategies) with the best performance in the in-sample window of data.[18] The returns to all strategies are simulated

---

[18]We denote this strategy as "combined" (COMBI) strategy and believe that this strategy mimics most closely the actual trader behavior. Specifically, the trader, that follows this strategy, using the in-sample window of data evaluates the performances of 24 MOM(n) strategies, 222 SMAC(s, l) strategies, 480 SMAE(n, p) strategies, and 2,442 EMAC(s, l, n) strategies; totally 3,168 strategies. The strategy with the best in-sample performance is used to generate the trading signal for the next period. In this combined strategy the trading rule may alter every each period. For example, one period the trader may use the MOM rule, next period the SMAE rule, and after that the SMAC rule. It is worth noting that, to the

accounting for 0.25% one-way transaction costs. In all strategies a Sell signal is a signal to leave the stocks and move to cash (or stay invested in cash). The performance of all strategies is measured using the Sharpe ratio. The out-of-sample returns are simulated from January 1870 to December 2015. The initial in-sample segment covers the period from January 1860 to December 1869.[19]

Table 9.9 reports the descriptive statistics of the buy-and-hold strategy over the out-of-sample period, as well as the descriptive statistics and performances of the moving average trading strategies simulated out-of-sample using both a rolling and an expanding in-sample window. The descriptive statistics include the (annualized) mean returns, the minimum and maximum monthly return. The following risk measures are reported: the (annualized) standard deviation of returns, the maximum drawdown,[20] the average maximum drawdown which is an equally-weighted average of the 10 largest drawdowns, and the average drawdown. The shape of the return distribution is characterized by skewness and kurtosis.[21] The outperformance is measured by $\Delta = SR_{MA} - SR_{BH}$ where SR_{MA} and SR_{BH} are the Sharpe ratios of the moving average strategy and the buy-and-hold strategy respectively. P-value is the value of testing the following null hypothesis $H_0 : \Delta \leq 0$. This hypothesis is tested using the stationary block-bootstrap method consisting in drawing 10,000 random resamples with the average block length of 5 months.

Figure 9.7 shows the rolling 10-year out-of-sample outperformance produced by the trading strategies simulated using both a rolling and an expanding in-sample window. Apparently, this figure clearly demonstrates that the

---

(Footnote 18 continued)
best knowledge of the author, in all previous studies the researchers tested the performance of a single rule at a time. For instance, one tested separately the performance of the MOM and SMAC rules. Such test method implicitly assumes that the trader always uses a single arbitrary rule; and there is absolutely no justification for why the trader has to follow a single rule.

[19]To check the robustness of findings reported in this section, we varied the length of the initial in-sample segment from 5 to 20 years. Qualitatively, the conclusion reached in this section remains intact regardless of the length of the initial in-sample segment.

[20]Drawdown is a measure of the decline from a historical peak to the subsequent trough. The amplitude of a drawdown is measured as $A = \frac{P_{peak} - P_{trough}}{P_{peak}}$, where P_{peak} is the stock price at a historical peak and P_{trough} is the stock price at the subsequent trough. The maximum drawdown is the maximum of all drawdowns over some given historical period. To compute all drawdown measures, using the time-series of total returns to a strategy we construct the series of prices. As a result, we compute the drawdowns using the prices adjusted for dividends.

[21]Skewness is a measure of the asymmetry of the probability distribution. Skewness can be both positive and negative. Negative (positive) skew indicates that the tail on the left (right) side of the probability distribution function is longer or fatter than that on the right (left) side. The skewness of the normal distribution equals to 0. Kurtosis is a measure of whether the probability distribution is heavy-tailed or light-tailed relative to the normal distribution. The kurtosis of the normal distribution equals to 3. Kurtosis above (below) 3 indicates that the probability distribution is heavy (light) tailed relative to the normal distribution.

Table 9.9 Descriptive statistics of the buy-and-hold strategy and the out-of-sample performance of the moving average trading strategy

	BH	ROL	EXP
Mean returns %	10.15	8.14	9.23
Std. deviation %	17.28	11.95	11.64
Minimum return %	−29.43	−23.51	−23.51
Maximum return %	42.91	42.66	42.91
Skewness	0.28	0.53	0.74
Kurtosis	8.86	17.34	19.72
Average drawdown %	7.25	5.89	5.32
Average max drawdown %	41.25	26.67	24.03
Maximum drawdown %	83.14	62.96	45.82
Outperformance		0.00	**0.10**
P-value		0.52	0.08
Rolling 5-year Win %		37.45	56.59
Rolling 10-year Win %		45.44	65.58

Notes BH denotes the buy-and-hold strategy. ROL denotes the moving average trading strategy simulated using a rolling in-sample window. EXP denotes the moving average trading strategy simulated using an expanding in-sample window. Mean returns and standard deviations are annualized. Bold text indicates the outperformance which is statistically significant at the 10% level

outperformance is not only very uneven over time, but often a moving average trading strategy underperforms its passive counterpart. Therefore the reported outperformance is a measure of average outperformance computed using a very long horizon (which is beyond the investment horizon of any individual investor). Since the majority of investors have short- to medium term horizons, the average outperformance produced by a moving average trading strategy over a horizon of 155 years is not especially relevant. In order to provide a more accurate picture of outperformance, using rolling windows of 5 and 10 years we compute the probability that the moving average trading strategy outperforms its passive counterpart over an arbitrary historical period of 5 and 10 years. These probabilities are denoted as "Rolling 5(10)-year Win %".

The conclusion that can be reached from the results reported in Table 9.9, coupled with the graphical illustration of rolling 10-year outperformance in Fig. 9.7, is pretty straightforward: the out-of-sample performance of the moving average trading strategy simulated using an expanding in-sample window is substantially better than that of its counterpart simulated with a rolling in-sample window.[22] Whereas the moving average strategy simulated using an

---

[22]To check the robustness of this finding, we also analyzed the out-of-sample performance of single trading rules. We found that only the MOM(n) rule showed better out-of-sample performance when the returns to this rule were simulated using a rolling in-sample window. However, the evidence of superior out-of-sample performance of this rule, simulated with a rolling in-sample window, appeared mainly during the first part of the historical sample.

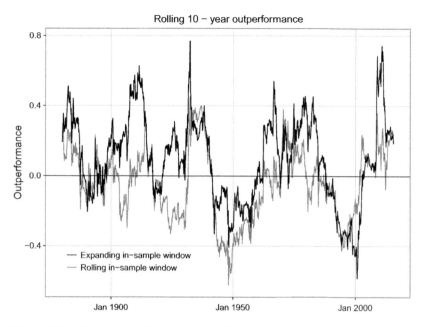

Fig. 9.7 Rolling 10-year out-of-sample outperformance produced by the trading strategies simulated using both a rolling and an expanding in-sample window. The out-of-sample segment cover the period from January 1870 to December 2015. The first point in the graph gives the outperformance over the first 10-year period from January 1870 to December 1879. Outperformance is measured by $\Delta = SR_{MA} - SR_{BH}$ where SR_{MA} and SR_{BH} are the Sharpe ratios of the moving average strategy and the buy-and-hold strategy respectively

expanding in-sample window both economically and statistically significantly (at the 8% level) outperforms the buy-and-hold strategy, the moving average strategy simulated using a rolling in-sample window has the same (risk-adjusted) performance as the buy-and-hold strategy. As compared with the strategy simulated using a rolling in-sample window, the strategy simulated using an expanding in-sample window has higher mean returns, lower riskiness, and higher probability of beating the passive strategy over short- to medium-term horizons.

Why the out-of-sample performance of a moving average trading strategy simulated using an expanding in-sample window is better than that of its counterpart simulated using a rolling window? This result seems to be counterintuitive taking into account the evidence that the best trading strategy in a back test varies over time. We propose several explanations for this result. First, when the sample size is relatively short, the data mining bias is large and, consequently, the performance of the best trading strategy in a back test has a large random component. Second, even if the variations in the type of the best trading strategy in a back test are not due to randomness alone, the market's dynamics may change way too fast. As a result of fast changing market's dynamics, trading rules that were optimal in the near past may no longer be optimal in the near future. Third, the advantage of the moving average trading strategy appears mainly during the periods of severe market downturns (for the motivation, see Fig. 9.7). During such periods, the optimal trading strategy may be more or less the same. That is, the trading strategy that was optimal during the decade of 1930s may again be optimal (or close to optimal) during the decades of 1970s and 2000s. We conjecture that the moving average strategies with a window size of 10–12 months (examples are the SMAC(2,10) and P-SMA(12) strategies) are the strategies that work best during the severe market downturns.

9.5.2 Ambiguity in Performance Measurement

Because both in-sample and out-of-sample performance of a moving average trading strategy is very uneven over time, the results of both in-sample and out-of-sample tests of profitability of moving average trading rules depend on the choice of the historical period where the trading rules are tested. In addition, the out-of-sample performance of trading rules depends, sometimes crucially, on the choice of split point between the initial in-sample and out-of-sample subsets. The goal of this section is to illustrate these issues.

To illustrate the dependence of the out-of-sample outperformance of the moving average trading strategy on the choice of split point, we select the historical period from January 1900 to December 2015 and simulate the out-of-sample returns to the moving average trading strategy using an expanding in-sample window. We vary the split point between the initial in-sample and out-of-sample segments from January 1910 to January 2011. Figure 9.8, upper panel, plots the out-of-sample outperformance of the moving average trading strategy for different choices of the sample split point. The lower panel of this figure plots the p-value of the test for outperformance.

Apparently, both the outperformance and the p-value of the test for outperformance depend significantly on the choice of split point. In particular, for the majority of choices, despite the fact that the p-value of the outperformance test is above 10%, the outperformance is positive. Further note that the trading strategy's outperformance increases dramatically if the split point is displaced towards the end of the sample. In particular, if the sample split point is located in between 1995 and 2005, the outperformance is more than double as high as that with the other choices for the sample split point. When the split point is located either from 1921 to 1930, or during the decade of 1990s, the p-value of the test is either below the 10% level or just a bit above

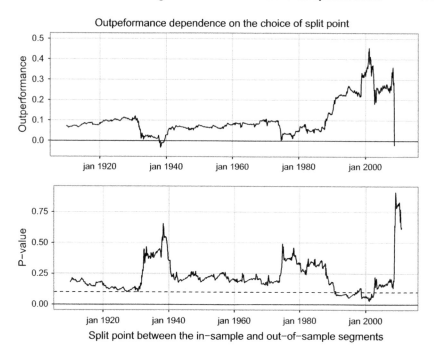

Fig. 9.8 *Upper panel* plots the out-of-sample outperformance of the moving average trading strategy for different choices of the sample split point. The outperformance is measured over the period that starts from the observation next to the split point and lasts to the end of the sample in December 2015. The *lower panel* of this figure plots the p-value of the test for outperformance. In particular, the following null hypothesis is tested: $H_0 : SR_{MA} - SR_{BH} \leq 0$ where SR_{MA} and SR_{BH} are the Sharpe ratios of the moving average strategy and the buy-and-hold strategy respectively. The *dashed horizontal line* in the *lower panel* depicts the location of the 10% significance level

this level. The outperformance is statistically significant at the 5% level when the split point is located from 1998 to 2001. Therefore, it is possible to choose the location of the split point such that the result of the out-of-sample test of profitability favors the moving average strategy and leads to the conclusion that the outperformance of the moving average trading strategy is positive and statistically significantly above zero at the conventional statistical levels (5% or 10%). Note, however, that for some "unfortunate" choices for the split point location, the out-of-sample outperformance is either close to zero or negative. Specifically, this is the case when split points are located either from 1930 to 1940 or from 1975 to 1980. If the split point belongs to either of the two specific periods, then one arrives at the opposite conclusion: the performance of the market timing strategy is either equal to or worse than that of the buy-and-hold strategy.

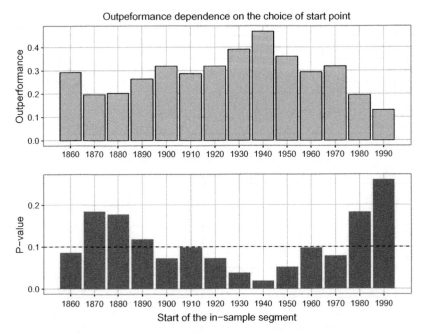

Fig. 9.9 *Upper panel* plots the out-of-sample outperformance of the moving aver-age trading strategy for different choices of the sample start point. Regardless of the sample start point, the out-of-sample segment covers the period from January 2000 to December 2015. The *lower panel* of this figure plots the p-value of the test for outperfor-mance. In particular, the following null hypothesis is tested: $H_0 : SR_{MA} - SR_{BH} \le 0$ where SR_{MA} and SR_{BH} are the Sharpe ratios of the moving average strategy and the buy-and-hold strategy respectively. The *dashed horizontal line* in the *lower panel* depicts the location of the 10% significance level

To illustrate the dependence of the out-of-sample outperformance of the moving average trading strategy on the choice of the historical period, we simulate the out-of-sample returns to the moving average trading strategy over the period from January 2000 to December 2015. We vary the start of the historical period from 1860 to 1990 with a step of 10 years. In other words, the start of the in-sample segment of data takes values in 1860, 1870, and so on up to 1990. Figure 9.9, upper panel, plots the out-of-sample outperformance of the moving average trading strategy for different choices of the sample start point. The lower panel of this figure plots the p-value of the test for outperformance.

Again, the graphs in this figure clearly illustrate that the out-of-sample outperformance of the moving average trading strategy depends very much on the sample start point. Despite the fact that the out-of-sample period from 2000 to 2015 was very successful for the market timing strategies (because this particular period contains two severe stock market crashes: the Dot-Com bubble crash of 2001–02 and the Global Financial Crisis of 2007–08) and the outperformance delivered by the moving average trading strategy is always positive regardless of the sample start point, the p-value of the outperformance test depends significantly on the choice of the sample start point. The best outperformance and the lowest p-value of the outperformance tests are attained when the sample start point is chosen as January 1940. If the sample starts either in 1870, 1880, 1980, or 1990, the p-value of the outperformance tests is way above the 10% level.

The illustrations provided in this section suggest that it is very difficult to provide an objective assessment of the historical outperformance delivered by the moving average trading strategy. This is because the outperformance depends on many different choices: the set of trading rules, the type of forward test (specifically, the choice of either expanding or rolling in-sample window), the choice of historical sample period, and the choice of the split point between the initial in-sample and out-of-sample segments. Therefore one needs to keep in mind this ambiguity in out-of-sample performance measurement. In the subsequent analysis, our choices for historical periods and split points are made in order to provide the most typical picture of the out-of-sample outperformance that is delivered by the moving average trading strategy.

9.5.3   Main Results of Forward Tests

In this section we report the detailed results of forward (that is, out-of-sample) tests of the moving average trading strategies. We forward-test some single trading rules and the combined rule where at each month-end we select the rule with the best performance in the in-sample segment of data. The following single rules are tested:

$MOM(n)$ for $n \in [2, 25]$, totally 24 trading strategies;

$P\text{-}SMA(n)$ for $n \in [2, 25]$, totally 24 trading strategies;

$SMAC(s, l)$ for $s \in [1, 12]$ and $l \in [2, 25]$, totally 222 trading strategies;

$SMAE(n, p)$ for $n \in [2, 25]$ and $p \in [0.25, 0.5, \ldots, 5.0]$, totally 480 trading strategies;

$EMACD(s, l, n)$ for $s \in [1, 12], l \in [2, 25]$, and $n \in [2, 12]$, totally 2,442 trading strategies.

The motivation for forward-testing the P-SMA rule is that in this rule, as well as in the MOM rule, there is only one single parameter: the size of the averaging window. Generally, the less the number of parameters in a trading rule, the less the number of tested strategies and, consequently, the less the data mining bias in the performance of the best trading strategy in a back test. Therefore the out-of-sample performance of the P-SMA rule might be potentially better than the performance of the SMAC rule which generalizes the P-SMA rule.

In the combined strategy, the performance of each single strategy in all tested rules is evaluated in the in-sample segment of data, and the trading signal at month-end equals the trading signal of the strategy with the best performance in the in-sample segment of data. In the combined strategy, the overall number of tested single trading strategies amounts to 3,192. The returns to all strategies are simulated accounting for 0.25% one-way transaction costs. In all strategies a Sell signal is a signal to leave the stocks and move to cash (or stay invested in cash). The performance of all strategies is measured using the Sharpe ratio. The forward test is implemented with an expanding in-sample window. The null hypothesis of no outperformance is tested using the stationary block-bootstrap method consisting in drawing 10,000 random resamples with the average block length of 5 months.

Table 9.10 reports the descriptive statistics of the buy-and-hold strategy and the out-of-sample performance of the moving average trading strategies. The performance is reported for the full out-of-sample period from January 1870 to December 2015 (with the initial in-sample segment from January 1860 to December 1869), for the first part of the out-of-sample period from January 1870 to December 1943 (with the initial in-sample segment from January 1860 to December 1869), and for the second part of the out-of-sample period from January 1944 to December 2015 (with the initial in-sample segment from January 1929 to December 1943). It is important to note that the two sub-periods have exactly the same number of bull-bear market phases. In particular, each of the two sub-periods has 21 bull and 20 bear markets.

Judging by (the sign of) the estimated outperformance, every single moving average strategy and the combined strategy outperforms the buy-and-hold strategy on the risk-adjusted basis. This observation applies equally to the outperformances over the whole period and the two sub-periods. Over the whole period, 3 out of 5 single strategies and the combined strategy statistically significantly outperform (at the 10% level) the buy-and-hold strategy. The performance of the P-SMA rule is statistically significantly better than that of the buy-and-hold strategy at the 5% level.

Over the first sub-period, only the performance of the MACD rule is statistically significantly better than that of the buy-and strategy. Even though the outperformance delivered by the MOM, P-SMA, and the combined rule is only marginally below the outperformance of the MACD rule, for these rules we cannot reject (at conventional statistical levels) the hypotheses that their performance is not better than the performance of the buy-and-hold strategy.

Table 9.10 Descriptive statistics of the buy-and-hold strategy and the out-of-sample performance of the moving average trading strategies

Statistics	BH	Moving average strategy					
		MOM	P-SMA	SMAC	SMAE	MACD	COMBI
Total period from 1870 to 2015							
Mean returns %	10.15	9.29	9.42	8.86	8.69	9.33	9.23
Std. deviation %	17.28	11.72	11.41	11.40	11.36	11.53	11.64
Minimum return %	−29.43	−23.51	−23.51	−23.51	−23.51	−23.51	−23.51
Maximum return %	42.91	42.66	16.09	16.09	16.09	42.91	42.91
Skewness	0.28	0.68	−0.49	−0.49	−0.42	0.76	0.74
Kurtosis	8.86	18.21	6.15	6.28	5.76	20.53	19.72
Average drawdown %	7.25	5.26	4.99	5.25	5.26	5.18	5.32
Average max drawdown %	41.25	22.21	23.30	23.02	24.63	23.71	24.03
Maximum drawdown %	83.14	47.01	51.65	44.50	53.46	44.01	45.82
Outperformance		0.10	0.13	0.08	0.07	**0.11**	0.10
P-value		0.06	0.04	0.14	0.19	0.06	0.08
Rolling 5-year Win %		48.91	50.50	38.69	45.42	56.88	56.59
Rolling 10-year Win %		52.05	61.24	50.52	47.40	66.26	65.58
First period from 1870 to 1943							
Mean returns %	8.66	8.02	7.83	7.05	6.76	8.71	8.52
Std. deviation %	19.68	12.48	11.80	11.78	11.60	13.03	13.03
Minimum return %	−29.43	−23.51	−23.51	−23.51	−23.51	−23.51	−23.51
Maximum return %	42.91	42.66	16.09	16.09	16.09	42.91	42.91
Skewness	0.56	1.44	−0.53	−0.49	−0.43	1.29	1.29
Kurtosis	9.67	25.44	7.40	7.58	7.08	22.55	22.53
Average drawdown %	9.16	6.47	6.00	6.53	6.73	6.49	6.77
Average max drawdown %	32.20	19.03	18.52	18.65	20.76	20.14	20.46
Maximum drawdown %	83.14	47.01	51.65	44.50	53.46	44.01	45.82
Outperformance		0.10	0.11	0.04	0.02	**0.14**	0.13
P-value		0.15	0.17	0.34	0.44	0.07	0.11
Rolling 5-year Win %		47.65	57.54	38.72	51.39	63.21	62.73
Rolling 10-year Win %		51.50	76.33	52.67	47.98	79.32	78.93
Second period from 1944 to 2015							
Mean returns %	11.69	10.15	10.52	10.37	10.69	9.54	10.41
Std. deviation %	14.40	11.03	10.86	11.04	11.04	9.63	10.94
Minimum return %	−21.54	−21.54	−21.54	−21.54	−21.54	−21.54	−21.54
Maximum return %	16.78	13.21	12.17	13.46	13.46	12.17	13.46
Skewness	−0.41	−0.47	−0.51	−0.47	−0.41	−0.54	−0.40
Kurtosis	1.60	4.46	4.52	4.42	4.20	7.25	4.35
Average drawdown %	5.99	4.79	4.47	4.55	4.37	4.32	4.60
Average max drawdown %	28.91	16.03	15.09	15.81	14.65	14.05	14.88
Maximum drawdown %	50.96	23.26	23.26	24.28	23.26	23.26	23.26
Outperformance		0.02	0.06	0.04	0.07	0.04	0.05
P-value		0.42	0.26	0.35	0.23	0.38	0.30
Rolling 5-year Win %		35.65	48.20	41.99	39.13	45.96	39.38
Rolling 10-year Win %		44.16	51.01	50.20	51.81	57.72	49.13

Notes **B**H denotes the buy-and-hold strategy, whereas COMBI denotes the "combined" moving average trading strategy where at each month-end the best trading strategy in a back test is selected. The notations for the other trading strategies are self-explanatory. Outperformance is measured by $\Delta = SR_{MA} - SR_{BH}$ where SR_{MA} and SR_{BH} are the Sharpe ratios of the moving average strategy and the buy-and-hold strategy respectively. Bold text indicates the outperformance which is statistically significant at the 10% level

However, this result can be explained by the fact that the statistical power of any test reduces with decreasing sample size.

Despite the fact that the two sub-periods have the same number of bull and bear markets, in the second sub-period the stock market has been much more often in the bull state. Therefore, as could be expected beforehand, over the second sub-period the moving average trading strategies outperformed the passive strategy to a much lesser extent. Specifically, whereas over the first sub-period the average outperformance (measured by $\Delta = SR_{MA} - SR_{BH}$) amounts to 0.090, over the second sub-period the average outperformance is reduced by half and amounts to 0.047. Similarly, while over the first sub-period the probability, that a moving average trading strategy outperforms its passive counterpart over a 10-year horizon, varies from 51% to 79%, over the second sub-period this probability is reduced and varies from 44% to 57%. Consequently, the advantage of the moving average trading strategy over the buy-and-hold strategy has diminished through time.

The comparison of the descriptive statistics of the returns to the moving average trading strategies versus the descriptive statistics of the returns to the buy-and-hold strategy reveals the following. Judging by the values of the standard deviation of returns (a.k.a. volatility), all moving average trading strategies are virtually equally risky. We observe a significant reduction in return volatility as compared to the volatility of the passive strategy. However, the reduction of volatility is not surprising because virtually in any moving average strategy about 1/3 of the time the money are held in cash. The mean returns to the moving average strategies are also below the mean returns to the passive strategy; the only exception is the mean return to the MACD rule over the first sub-period. Thus, the moving average trading strategy has both lower returns and risk as compared to those of its passive counterpart. Consequently, over the long run the cumulative return to the buy-and-hold strategy tends to increase faster than the cumulative return to the moving average strategy. Figure 9.10, upper panel, demonstrates this feature by plotting the cumulative returns to the buy-and-hold strategy and the out-of-sample cumulative returns to the P-SMA rule.

The advantages of the moving average trading strategy are more pronounced when one compares the drawdown-based measures of risk of the moving average strategy and the corresponding buy-and-hold strategy. Over the total sample period, whereas the reduction of volatility amounts to approximately 1/3, the reduction of the maximum drawdown and the average maximum drawdown amounts to approximately 1/2. Thus, and it is very important to emphasize, the moving average trading strategy is not a "high returns, low risk" strategy as compared to the buy-and-hold strategy. In reality, it is a "low returns,

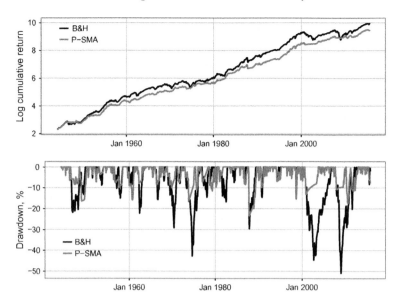

Fig. 9.10 *Upper panel* plots the cumulative returns to the P-SMA strategy versus the cumulative returns to the buy-and-hold strategy (B&H) over the out-of-sample period from January 1944 to December 2015. *Lower panel* plots the drawdowns to the P-SMA strategy versus the drawdowns to the buy-and-hold strategy over the out-of-sample period

low risk" strategy. However, for all trading rules the decrease in mean (excess) return is smaller than the decrease in volatility. This property improves the risk-adjusted performance of a moving average strategy as compared with that of the passive strategy. Most importantly, for all trading rules the decrease in mean (excess) return is much smaller than the decrease in drawdown-based measures of risk. Therefore the main advantage of the moving average trading strategy lies in its superior downside protection. Figure 9.10, lower panel, demonstrates this advantage by plotting the drawdowns to the buy-and-hold strategy and the P-SMA rule. We will elaborate more on this property of the moving average trading strategy at the end of this chapter.

9.5.4   Performance over Bull and Bear Markets

To gain further insights into the properties of the moving average trading strategy, we analyze the out-of-sample performance of the combined moving average trading strategy and the performance of the corresponding buy-and-hold strategy over bull and bear markets. We focus on the second part of the out-of-sample period, from January 1944 to December 2015, because, in our opinion, the performance over this particular historical period can be used as a reliable estimate of the expected future performance. Table 9.11 reports the descriptive statistics of the buy-and-hold strategy and the moving average trading strategy over bull and bear markets. The descriptive statistics include the mean and standard deviation of returns (in annualized terms), as well as the Sharpe ratios over the bull markets. The Sharpe ratios over the bear markets are not reported, because when the mean excess return is negative, the value of the Sharpe ratio is not reliable and hard to interpret. Figure 9.11 visualizes the mean returns and standard deviations of the moving average trading strategy and the corresponding buy-and-hold strategy over bull and bear markets.

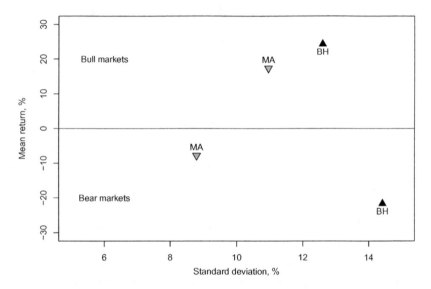

Fig. 9.11 Mean returns and standard deviations of the buy-and-hold strategy and the moving average trading strategy over bull and bear markets. **BH** and **MA** denote the buy-and-hold strategy and the moving average trading strategy respectively

Table 9.11 Descriptive statistics of the buy-and-hold strategy and the moving average trading strategy over bull and bear markets

| Statistics | Bull markets | | Bear markets | |
	BH	MA	BH	MA
Mean returns %	24.35	17.35	−21.61	−7.79
Std. deviation %	12.60	10.95	14.42	8.80
Sharpe ratio	1.63	1.24		

Notes **BH** and **MA** denote the buy-and-hold strategy and the moving average trading strategy respectively. Mean returns and standard deviations are annualized. Descriptive statistics are reported for the out-of-sample period from January 1944 to December 2015

Apparently, over bull markets the buy-and-hold strategy outperforms the moving average trading strategy. Specifically, over bull markets the buy-and-hold strategy delivers both higher mean returns and higher Sharpe ratio than the moving average trading strategy. It is interesting to observe that the moving average trading strategy has lower standard deviation of returns (as compared to that of the buy-and-hold strategy) over both bull and bear markets. Specifically, as compared with the standard deviation of returns to the buy-and-hold strategy, the standard deviation of returns to the moving average trading strategy is less by 13% (39%) over the bull (bear) markets. That is, over bull markets the buy-and-hold strategy has higher returns and higher risk than the moving average strategy, but this strategy has better risk-adjusted performance than the moving average strategy. On the other hand, over bear markets the moving average trading strategy has better tradeoff between the risk and return than that of the buy-and-hold strategy. In particular, over bear markets the moving average strategy is "high returns, low risk" strategy. It is worth noting, however, that in bear markets the mean returns to the moving average strategy are negative (nevertheless they are much higher than the mean returns to the buy-and-hold strategy in bear markets). That is, on average, technical traders who employ the moving average strategy also lose money in bear markets; yet their losses are less than those of the investors that follow the buy-and-hold strategy.

To help explain the results presented in Table 9.11, we analyze the similarity, or concordance, between the Bull-Bear states of the market and the Buy-Sell periods produced by the trading signals generated by the (out-of-sample) moving average trading strategy. Figure 9.12 visualizes the Bull-Bear states of the market and the Buy-Sell periods. Obviously, the similarity is far from perfect. There are many Sell signals during Bull market states, as well as there are many Buy signals during Bear market states. The reader is reminded that one of the essential properties of moving averages is that they detect a change in the stock market trend with some delay. By examining the plot in Fig. 9.12, one can easily note that, roughly, the Buy-Sell periods represent delayed copies of the Bull-Bear market states.

Our analysis reveals that the number of distinctive Buy-Sell periods is approximately double as high as the number of corresponding Bull-Bear stock market states. For example, over the tested period from 1944 to 2015, there

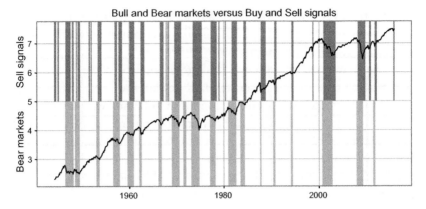

Fig. 9.12 Bull and Bear markets versus Buy and Sell signals generated by the moving average trading strategy. *Shaded ares* in the *upper part* of the plot indicate Sell periods. *Shaded areas* in the *lower part* of the plot indicate Bear market states

were 21 Bull markets and 37 Buy periods. To quantify the similarity between the Bull-Bear states of the market and the Buy-Sell periods, we employ the Simple Matching Coefficient (SMC). Denoting by $Signal_t$ the trading signal for month t and by $State_t$ the state of the market in month t, the computation of the SMC starts with calculating the following quantities:

M_{00} = the number of instances where $Signal_t$ = Sell and $State_t$ = Bear,

M_{01} = the number of instances where $Signal_t$ = Sell and $State_t$ = Bull,

M_{10} = the number of instances where $Signal_t$ = Buy and $State_t$ = Bear,

M_{11} = the number of instances where $Signal_t$ = Buy and $State_t$ = Bull.

Notice that M_{00} and M_{11} can be interpreted as the number of months with correct Sell and Buy signals respectively. In contrast, M_{01} and M_{10} can be interpreted as the number of months with false Sell and Buy signals respectively. For any month $t \in [1, T]$, each instance must fall into one of these four categories, meaning that

$$M_{00} + M_{01} + M_{10} + M_{11} = T.$$

The Simple Matching Coefficient is computed as the number of months with correct Buy and Sell trading signals divided by the total number of months

$$SMC = \frac{M_{00} + M_{11}}{M_{00} + M_{01} + M_{10} + M_{11}}.$$

The value of the SMC is constrained to lie within the range [0, 1], where the case $SMC = 1$ (or 100%) indicates a perfect match between the Bull-Bear market states and the Buy-Sell periods. Therefore the closer the similarity coefficient to unity, the better the moving average trading strategy identifies the stock market states. The computed value of the SMC of the moving average strategy equals 0.764. This value means that, over the tested period, the accuracy of this strategy was 76.4%. In other words, the moving average rules produced correct trading signals approximately 3/4 of time. Since the value of the SMC is substantially below 100%, we can conclude that the moving average trading strategy generates many false signals. The buy-and-hold strategy can be considered as a strategy which correctly (incorrectly) identifies all bull (bear) markets. The accuracy of the buy-and-hold strategy, as measured by the SMC, amounts to 72.5%.[23] Therefore, the moving average strategy is, in principle, just a bit more accurate than the buy-and-hold strategy in identification of the stock market states. Nevertheless, this very marginal increase in accuracy translates into a substantial downside protection.

To estimate the average lag time between the Buy-Sell periods and the Bull-Bear states, we back-shift the time series of the Buy-Sell periods, and for each lagged time-series of Buy-Sell periods, we compute the SMC. The average lag time is found as the number of back-shifts at which the SMC attains maximum. Formally, the average lag time is estimated as

$$\text{Average lag time} = \arg\max_{k \geq 1} \quad SMC(\text{State}_t, L^k \text{Signal}_t),$$

where L^k is the lag (or back-shift) operator defined by

$$L^k \text{Signal}_t = \text{Signal}_{t-k}.$$

The computation of the average lag time in the identification of the stock market states gives 4 months. That is, on average, the moving average trading strategy recognizes the change in the stock price trend with a lag time of 4 months (in out-of-sample tests). This result suggests that, in order the moving average trading strategy to work, the duration of the stock market state should be substantially longer than 4 months. Since over the second part of the sample the median duration of a Bear market was 12 months, we can roughly estimate that the moving average trading strategy works every second bear market on

---

[23]This number also tells us that over the second period the market was in Bull state 72.5% of time.

average. That is, roughly, the moving average trading strategy works (does not work) when the duration of a bear market is longer (shorter) than 12 months.

It is worth mentioning that the actual lag time in identification of a particular state of the market can deviate substantially from the average lag time. Consider, as an illustrative example, a concrete bear market that lasted only 3 months: from September 1987 to November 1987. This period includes the famous stock market crash that happened on October 19, 1987. Because the drop in the stock market prices during October 1987 was sharp and significant, the moving average trading strategy generated a Sell signal already for November 1987. That is, in this example, the lag time in the identification of the Bear market was only 2 months. Again, because during this Bear market the stock prices decreased swiftly and substantially, the value of the moving average was higher than the stock prices during a long period after the beginning of the subsequent Bull market. The moving average trading strategy recognized this Bull market with a delay of 11 months.

9.6   Daily Trading the S&P Composite Index

The goal of this section is to find out whether there is any advantage in trading using the daily data versus the monthly data. In principle, the daily data are freely available and it seems natural to expect that using the daily data may potentially improve the performance of the moving average strategy. This is because the high-frequency data are supposed to provide earlier Buy and Sell trading signals. To the best knowledge of the author, so far there is only a single paper by Clare et al. (2013) where the authors use daily and monthly data on the S&P 500 index (over the period from 1988 to 2011) and investigate this question using the back-testing methodology. Rather surprisingly, Clare et al. (2013) found that there is no advantage in trading daily rather than monthly. We re-examine this question using a longer sample of data, a larger set of trading rules, and both the back-testing and forward-testing methodology.

9.6.1   Data

Daily prices on the S&P Composite stock market index are obtained from the Center for Research in Security Prices (CRSP).[24] The data span the period from July 2, 1926 to December 31, 2015. Dividends are 12-month moving sums of dividends paid on the Standard and Poor's Composite index. The

---

[24]http://www.crsp.com/.

monthly dividend series data are provided by Amit Goyal.[25] The monthly risk-free rate of return data for the sample period are obtained from the data library of Kenneth French.[26] This rate equals to 1-month Treasury Bill rate from Ibbotson and Associates Inc.

Until the end of 1952, stock exchanges in the US were open 6 days a week. Beginning from 1953, stocks were traded 5 days a week only. Therefore, for the sake of consistency of daily data series, we remove the index values for Saturdays. Daily index values are used to compute the daily capital gain returns. The daily dividend yield is the simple daily yield that, over the number of trading days in the month, compounds to 1-month dividend yield. The total returns are obtained by summing up the capital gain returns and the dividend yields. The daily risk-free rate is the simple daily rate that, over the number of trading days in a given month, compounds to 1-month Treasury Bill rate from Ibbotson and Associates Inc.

9.6.2 Back-Testing Trading Rules

The following set of rules are tested:

MOM(n)   for $n \in [2, 3, \ldots, 15, 20, 30, \ldots, 350]$, totally 48 trading strategies;

SMAC(s, l)   for $s \in [1, 2, \ldots, 20, 25, \ldots, 80]$ and $l \in [2, 3, \ldots, 15, 20, 30, \ldots, 350]$, totally 1,144 trading strategies;

SMAE(n, p)   for $n \in [2, 3, \ldots, 15, 20, 30, \ldots, 350]$ and $p \in [0.25, 0.5, \ldots, 5.0]$, totally 960 trading strategies;

EMACD(s, l, n)   for $s \in [1, 2, \ldots, 20, 25, \ldots, 80], l \in [2, 3, \ldots, 15, 20, 30, \ldots, 350]$, and $n \in [5, 10, \ldots, 20, 40, \ldots, 100]$, totally 9,152 trading strategies.

The overall number of tested trading strategies amounts to 11,304. The returns to all strategies are simulated accounting for 0.25% one-way transaction costs. In all strategies a Sell signal is a signal to leave the stocks and move to cash (or stay invested in cash). The performance of all strategies is measured using the Sharpe ratio.

Table 9.12 reports the top 10 best trading strategies in a back test over the period from January 1944 to December 2015, which corresponds to the second part of our sample of monthly data. The results reveal that the trading strategies

---

[25] http://www.hec.unil.ch/agoyal/.

[26] http://mba.tuck.dartmouth.edu/pages/faculty/ken.french/data_library.html.

Table 9.12 Top 10 best trading strategies in a back test over January 1944 to December 2015

Rank	Strategy	Δ
1	SMAE(230,2.5)	0.19
2	SMAE(230,2.75)	0.19
3	SMAE(220,2.75)	0.19
4	SMAE(160,4)	0.18
5	SMAE(210,3)	0.18
6	SMAE(210,3.25)	0.18
7	SMAE(220,3)	0.18
8	SMAE(170,3.75)	0.18
9	SMAE(190,3)	0.18
10	SMAE(180,4)	0.17

Notes $\Delta = SR_{MA} - SR_{BH}$ where SR_{MA} and SR_{BH} are the Sharpe ratios of the moving average strategy and the buy-and-hold strategy respectively

based on the SMAE(n, p) rule show the best performance in a back test. The corresponding results for back-tests using the monthly data are reported in Table 9.8. The comparison of the results for the monthly and daily data suggests the following two noteworthy observations. The first observation is that when monthly data are used, the best trading strategies are based on the SMAC(s, l) rule; the strategies based on the SMAE(n, p) rule perform only marginally worse than the strategies based on the SMAC(s, l) rule. In contrast, when daily data are used, the best trading strategies are based solely on the SMAE(n, p) rule; the strategies based on the SMAC(s, l) rule perform notable worse than the strategies based on the SMAE(n, p) rule. The second observation is that using daily data produces a higher outperformance in a back test than using monthly data. Specifically, whereas the best trading strategy in a back test outperforms the buy-and-hold strategy by $\Delta = 0.19$ when daily data are used, the best trading strategy in a back test outperforms the buy-and-hold strategy by $\Delta = 0.15$ when monthly data are used.

It is worth mentioning that, when daily data used, the most popular (among practitioners) moving average trading strategy is the SMAC(50,200). In order to test the robustness of our finding on the superior performance of the SMAE rule, we used several other choices for the test period and transaction cost and, regardless of the chosen sample period and amount of transaction costs (in 0.1–0.5% range), our results suggest that the SMAE rule always outperforms the SMAC rule. Given this fact, the broad popularity of the SMAC(50,200) strategy is rather surprising. To highlight the differences between the performances of the SMAC(50,200) strategy and the

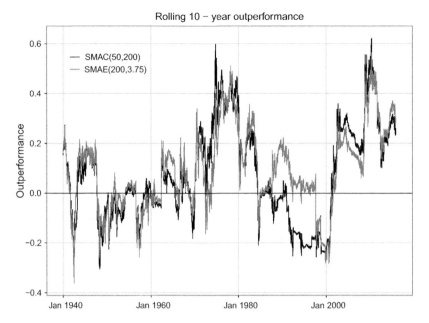

Fig. 9.13 Rolling 10-year outperformance produced by the SMAE(200,3.75) strategy and the SMAC(50,200) strategy over period from January 1930 to December 2015. The first point in the graph gives the outperformance over the first 10-year period from January 1930 to December 1939. Outperformance is measured by $\Delta = SR_{MA} - SR_{BH}$ where SR_{MA} and SR_{BH} are the Sharpe ratios of the moving average strategy and the buy-and-hold strategy respectively

SMAE(200,3.75) strategy,[27] Fig. 9.13 plots the rolling 10-year outperformance produced by the SMAE(200,3.75) strategy and the SMAC(50,200) strategy over period from January 1930 to December 2015. A visual comparison suggests that the performances of these two alternative strategies differ marginally. Only during the period from the mid-1980s to the mid-1990s the performance of the SMAC(50,200) strategy was significantly worse than that of the SMAE(200,3.75) strategy.

The reader is reminding that both the SMAC and SMAE rules generalize the P-SMA rule. Specifically, both the SMAC and SMAE rules are designed to reduce the number of whipsaw trades in the P-SMA rule. In this regard our results suggest that, when daily data are used, the best method of reducing the whipsaw trades is using a moving average envelope, not using a shorter moving average instead of the last closing price.

As a final but important remark, in our tests we always take into account transaction costs. The results on the best trading strategy in a back test in the

---

[27]The SMAE(200,3.75) is not the best trading strategy in a back test. We select this strategy because both the SMAC(50,200) and the SMAE(200,3.75) strategy use the same 200-day window to detect the trend.

absence of transaction costs are completely different. In particular, without transaction costs the best trading strategy in a back test over January 1944 to December 2015 is the MOM(2) strategy. Note that a Buy (Sell) trading signal in the MOM(2) strategy is generated when the close price for a day is higher (lower) than the close price the day before. Therefore this strategy consists in buying the stocks when the daily price change is positive, and selling the stocks otherwise; this strategy exploits a very short-term momentum. Figure 9.14 plots the rolling performance of this strategy over the period from January 1927 to December 2015. The graph in this plot suggests that the outperformance of this strategy was positive and increasing over the period from the early 1940s to the early 1970s. Afterwards, the outperformance delivered by the MOM(2) strategy was decreasing. From about the early 2000s the MOM(2) strategy started to underperform the buy-and-hold strategy. This fact suggests that the very short-term momentum in daily stock prices ceased to exist and was replaced by a very short-term mean-reversion.

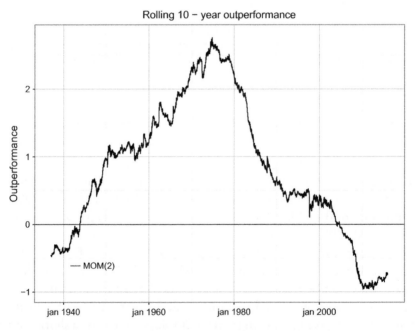

Fig. 9.14 Rolling 10-year outperformance produced by the MOM(2) strategy in the absence of transaction costs over the period from January 1927 to December 2015. Outperformance is measured by $\Delta = SR_{MA} - SR_{BH}$ where SR_{MA} and SR_{BH} are the Sharpe ratios of the moving average strategy and the buy-and-hold strategy respectively

9.6.3 Forward-Testing Trading Rules

The set of tested trading rules is the same as in the preceding section. To this set we add the P-SMA(n) rule where $n \in [2, 3, \ldots, 15, 20, 30, \ldots, 350]$. The forward-testing methodology is the same as for forward-testing trading rules using monthly data. The initial in-sample period is from January 1929 to December 1943. Consequently, the out-of-sample period is from January 1944 to December 2015. The forward test is implemented with an expanding in-sample window. To speed up the simulation of the out-of-sample strategy, the selection of the best trading strategy in the in-sample window is repeated every 21th day.

Table 9.13 reports the descriptive statistics of the buy-and-hold strategy and the out-of-sample performance of the moving average trading strategies when daily data are used. The corresponding results for monthly data are reported in Table 9.10. Rather surprisingly, among all single rules only the SMAE rule outperforms the buy-and-hold strategy in the out-of-sample test. Yet, there is no evidence that the SMAE rule statistically significantly outperforms the buy-and-hold strategy. The combined strategy also outperforms the buy-and-hold strategy; we guess that the combined strategy is largely based on using the SMAE rule. Other interesting observations that deserve our attention are as follows. First, the out-of-sample performance of the SMAC rule is economically

Table 9.13 Descriptive statistics of the buy-and-hold strategy and the out-of-sample performance of the moving average trading strategies

| Statistics | BH | Moving average strategy | | | | | |
		MOM	P-SMA	SMAC	SMAE	MACD	COMBI
Mean returns %	11.80	8.68	9.02	8.16	9.47	6.08	9.48
Std. deviation %	15.31	10.80	10.31	10.45	10.26	9.15	10.26
Minimum return %	−20.45	−6.86	−6.86	−20.45	−6.86	−6.86	−6.86
Maximum return %	11.59	5.12	5.12	5.12	5.12	5.12	5.12
Skewness	−0.66	−0.63	−0.45	−2.12	−0.52	−0.52	−0.52
Kurtosis	19.91	8.15	7.23	58.38	8.02	10.97	8.02
Average drawdown %	2.15	2.20	2.06	2.08	2.03	2.16	2.03
Average max drawdown %	32.60	21.16	17.07	18.90	16.69	18.80	16.23
Maximum drawdown %	55.23	37.59	24.48	39.79	22.59	29.65	21.48
Outperformance		−0.07	−0.02	−0.11	0.03	−0.28	0.03
P-value		0.80	0.58	0.87	0.41	0.99	0.39
Rolling 5-year Win %		39.06	38.96	22.30	40.86	24.34	40.86
Rolling 10-year Win %		36.46	46.81	23.69	47.03	19.83	47.03

Notes **BH** denotes the buy-and-hold strategy, whereas **COMBI** denotes the "combined" moving average trading strategy where at each month-end the best trading strategy in a back test is selected. The notations for the other trading strategies are self-explanatory. Outperformance is measured by $\Delta = SR_{MA} - SR_{BH}$ where SR_{MA} and SR_{BH} are the Sharpe ratios of the moving average strategy and the buy-and-hold strategy respectively

significantly below that of the P-SMA rule. In other words, our tests suggest that the SMAC rule is worse than the P-SMA rule in out-of-sample tests when daily data are used. Second, the performance of the EMACD rule is much worse than the performance of the buy-and-hold strategy. Third, with daily trading the out-of-sample outperformance is worse than that with monthly trading. Overall, our forward tests suggest that there is a disadvantage in trading daily rather than monthly. This result is probably counter-intuitive, but it strengthens the findings reported by Clare et al. (2013).

The natural question to ask is why daily trading is disadvantageous. We believe that the answer to this question lies in the fact that daily data are much noisier than monthly data. To illustrate the aforesaid, an engineering concept of "signal-to-noise" ratio can be used. In engineering, the signal-to-noise ratio is defined as the ratio of the signal power to the noise power. In technical analysis, the signal-to-noise ratio is sometimes used to measure the strength of a stock price trend. In this context, a signal-to-noise ratio can be computed as the (absolute) price change over some given period divided by a measure of price variability during the same period. The total price change in a given period can be expressed in terms of the mean price change or mean return; the price variability can be measured using the standard deviation of returns. Therefore, the daily and monthly signal-to-noise ratios can be measured by

$$\text{SNR}_d = \frac{\mu_d}{\sigma_d}, \quad \text{SNR}_m = \frac{\mu_m}{\sigma_m},$$

where SNR_d and SNR_m are the daily and monthly signal-to-noise ratios respectively, μ_d and σ_d are the daily mean return and standard deviation respectively, and μ_m and σ_m are the monthly mean return and standard deviation respectively. Since both the mean return and variance of returns are directly proportional to time, and there are approximately 21 trading days in a month, the relation between the monthly and daily signal-to-noise ratios are given by

$$\text{SNR}_m \approx \sqrt{21} \times \text{SNR}_d.$$

This means that the monthly signal-to-noise ratio is almost five times stronger than the daily signal-to-noise ratio. Therefore, it is easier to distinguish the signal from the noise when monthly data are used.

Given the fact that daily data are much noisier than monthly data, the random component of the observed outperformance of the best rule in a back test is greater when daily data are used. In other words, using daily data increases the data mining bias as compared with using monthly data. Therefore using daily data substantially increases the chances that the best trading rule in

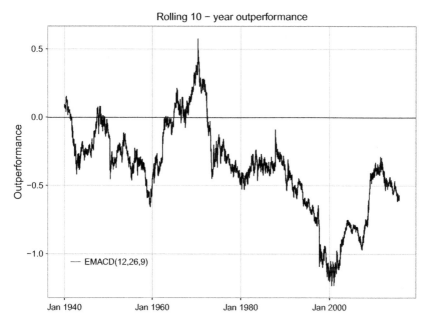

Fig. 9.15 Rolling 10-year outperformance produced by the EMACD(12,29,9) strategy over the period from January 1930 to December 2015. The first point in the graph gives the outperformance over the first 10-year period from January 1930 to December 1939. Outperformance is measured by $\Delta = SR_{MA} - SR_{BH}$ where SR_{MA} and SR_{BH} are the Sharpe ratios of the moving average strategy and the buy-and-hold strategy respectively

a back test is the rule that benefited most from good luck. The data mining bias increases dramatically when daily data are used, the sample size is rather short, and the computation of the trading signal in a technical trading rule depends on many parameters. Under these conditions, the best trading strategy in a back test usually performs very poorly out-of-sample, because the parameters of the trading strategy have been overfit to the in-sample data, a situation known as "backtest overfitting".[28]

In order to demonstrate the danger of backtest overfitting, consider the performance of the Moving Average Convergence/Divergence rule proposed by Gerald Appel (see Appel 2005) in the late 1970s. We remind the reader that the MACD rule uses three exponential moving averages (that is, the rule has three parameters) and Gerald Appel advocates that the best combination is to use moving averages of 12, 29, and 9 days. Figure 9.15 plots the rolling 10-year outperformance produced by the EMACD(12,29,9) strategy over the period from January 1930 to December 2015. The graph of the outperformance

---

[28] Overfitting is a concept borrowed from statistical regression analysis and machine learning. Overfitting denotes a situation when one fits a larger model than that required to capture the dynamics of the data. For more information on overfitting concept, see https://en.wikipedia.org/wiki/Overfitting.

reveals that the EMACD(12,29,9) strategy outperformed the buy-and-hold strategy basically only during a relatively short historical period from about the late 1960s to the late 1970s. Apparently, Gerald Appel "discovered" this strategy in the late 1970s by back-testing many different combinations of three moving averages using a sample of daily data of about 10 years long. Figure 9.15 convincingly demonstrates that neither before nor after the decade of 1970s the EMACD(12,29,9) strategy outperformed the buy-and-hold strategy. That is, the superior performance of the EMACD(12,29,9) strategy is a fluke, not a regular thing. It is unbelievable that still today, almost 40 years after the superior performance of this strategy was observed for the last time, numerous handbooks on technical analysis and numerous web-sites present the EMACD rule as "the most popular technical indicators in trading" and recommend using the EMACD(12,29,9) strategy for beating the market on a daily basis.

To recap, since daily data are much noisier than monthly data, the daily signal-to-noise ratio is much smaller than the monthly one; this feature makes the detection of a stock price trend more complicated with daily data. Using monthly data instead of daily allows one to effectively increase the signal-to-noise ratio and make easier to distinguish the signal from the noise.

9.7   Defending the Advantages of the Moving Average Strategy

9.7.1   The Use and Misuse of the Sharpe Ratio

The goal of this section is to elaborate in details on the precise meaning of the Sharpe ratio and any other rational reward-to-risk measure. The problem is that the Sharpe ratio seems to be a simple concept, but in practical applications the use of the Sharpe ratio is tricky. For example, the majority of practitioners fail to understand that the use of the Sharpe ratio is justified if the investor' preferences can be represented by a mean-variance utility function.[29] On the other hand, the majority of students who take an MBA degree (or a similar postgraduate degree) do know that the Sharpe ratio is related to the mean-variance utility function, but after taking investment courses all they remember is that the investor must select a portfolio with the highest Sharpe ratio. The students forget that the ultimate goal of the investor is to maximize the expected utility

---

[29]In addition, the majority of practitioner fail to understand that a rational performance measure is not any arbitrary ratio of reward to risk; a rational performance measure must satisfy a set of specific properties, see Cherny and Madan (2009) and Zakamulin (2010).

of his final wealth; in order to achieve this goal, the investor has to allocate optimally between the risk-free asset and the optimal risky portfolio.

To illustrate the misuse of the Sharpe ratio, and to highlight the fact that a portfolio with the highest Sharpe ratio can be inferior compared to another portfolio with a lower Sharpe ratio, consider the following problem presented to the students on the final exam in a postgraduate course on investments. In this problem the investor's attitude toward risk is represented by the mean-variance utility function defined over returns r

$$U(r) = E[r] - \frac{1}{200} A \times Var[r], \tag{9.1}$$

where the mean return and standard deviation are measured in percentages and the investor's coefficient of risk aversion $A = 2$. Initially, 70% of the investor's wealth is invested to stock A and the rest, 30%, is invested in the risk-free government securities. The mean return and standard deviation of returns of stock A are 10% and 31% respectively; the risk-free government securities provide the rate of return of 3%. The first question asks the students to compute the mean return, standard deviation, and Sharpe ratio of the investor's portfolio.

The problem continues as follows: The investor considers selling government securities and investing the proceeds in stock B. The mean return and standard deviation of returns of stock B are 12% and 36% respectively, and the correlation coefficient between returns to stocks A and B is 90%. The second question asks the students to compute the mean return, standard deviation, and Sharpe ratio of the portfolio of stocks A and B. The third and final question in this problem asks the students whether the investor should transfer money from the government securities to stock B.

Practically all students answer correctly to the first and second questions. Specifically, the correct answers are as follows (to save the space, we skip the computations). The mean return, standard deviation, and Sharpe ratio of the investor's original portfolio are 7.9%, 21.7%, and 0.226. The mean return, standard deviation, and Sharpe ratio of the portfolio of stocks A and B are 10.6%, 31.77%, and 0.239. Yet, only about 10% of students answer correctly to the last question. In particular, 90% of students use the Sharpe ratio as a decision criterion and reason as follows: since the Sharpe ratio of the portfolio of stocks A and B is higher than that of the portfolio of stock A and government securities (0.239 > 0.226), the investor should sell the government securities and invest the proceeds in stock B. This answer is incorrect because for this specific investor the (expected) utility from holding the portfolio of stocks

A and B is much lower than the utility of the initial portfolio of stock A and government securities. Indeed, the utility from the 70/30 portfolio of stock A and the risk-free securities

$$U(r) = 7.9 - \frac{2}{200} \times 21.7^2 = 3.19,$$

whereas the utility from the 70/30 portfolio of stock A and stock B

$$U(r) = 10.6 - \frac{2}{200} \times 31.77^2 = 0.51.$$

As a result, by reallocating money from the government securities to stock B, the investor significantly deteriorates his utility. Thus, this example demonstrates that a portfolio with the highest Sharpe ratio is not necessarily the portfolio that maximizes the investor's utility.

The reader is reminded that even though the utility function given by (9.1) is defined over returns, in reality it is a simplified form of the utility function defined over the investor's final wealth, see Chap. 7. The investor's ultimate goal is not to maximize the Sharpe ratio of his portfolio, but to maximize the utility that can be derived from his final wealth. According to modern finance theory, in order to maximize the utility the investor has to solve two interrelated problems: (1) select the optimal risky portfolio and (2) select the optimal capital allocation between the risk-free asset and the (optimal) risky portfolio. The Sharpe ratio allows the investor to solve only one problem: to select the optimal risky portfolio. However, the ultimate investor's goal is not fulfilled unless the investor selects the optimal capital allocation. Unfortunately, modern finance theory gives very little consideration to the solution of the second investor's problem. All modern finance theory says is that the optimal capital allocation depends on the investor's coefficient of risk aversion A; the investor needs to know the value of his A and make the optimal capital allocation according to his A. Overall, modern finance theory is basically oriented towards the needs of a portfolio manager (that is, it tells how to construct the optimal risky portfolio), not towards the needs of investors (it does not give practical advice on how to optimally allocate money between the risk-free asset and the risky portfolio). Therefore for practical investor's needs the use of the Sharpe ratio makes little sense if the investor does not know how to allocate money optimally between the risky portfolio and the risk-free asset.

Another important thing to remember is that the arguments behind the use of the Sharpe ratio assume the existence of a risk-free asset. These arguments break down in the absence of the risk-free asset. That is, the Sharpe ratio can be justified only when the risk-free asset is present. If there is no risk-free asset, then modern finance theory tells that the choice of the optimal risky portfolio is not unique; in this case the optimal risky portfolio depends on the investor risk preferences (that is, on the investor's coefficient of risk aversion). To make the further exposition more concrete, assume that the investor considers the choice between investing either in portfolio A or portfolio B. Denote the mean return and standard deviation of portfolio A by μ_A and σ_A respectively, and the mean return and standard deviation of portfolio B by μ_B and σ_B respectively.

In some cases the choice of the best risky portfolio does not depend on the investor's coefficient of risk aversion. Specifically, according to the mean-variance criterion, portfolio A dominates portfolio B if

$$\mu_A \geq \mu_B \text{ and } \sigma_A \leq \sigma_B$$

and at least one inequality is strict. To see this, consider the investor utilities

$$U(r_A) = \mu_A - \frac{1}{2}A\sigma_A^2 \text{ and } U(r_B) = \mu_B - \frac{1}{2}A\sigma_B^2.$$

Let us find the difference between $U(r_A)$ and $U(r_B)$

$$U(r_A) - U(r_B) = (\mu_A - \mu_B) - \frac{1}{2}A\left(\sigma_A^2 - \sigma_B^2\right).$$

Since $\mu_A - \mu_B \geq 0$ and $\sigma_A^2 - \sigma_B^2 \leq 0$ and at least one inequality is strict, we conclude that

$$U(r_A) - U(r_B) > 0.$$

That is, regardless of the value of risk aversion coefficient A, the utility of portfolio A is higher than that of portfolio B. Consequently, the choice of the best risky portfolio is easy when some portfolio has higher mean return (i.e., reward) and, at the same time, lower standard deviation (i.e., risk) than the other portfolio. In this situation, portfolio A has higher reward and lower risk than those of portfolio B.

Consider another, much more typical situation:

$$\mu_A > \mu_B \text{ and } \sigma_A > \sigma_B.$$

That is, in this case portfolio A has higher mean return and higher risk than portfolio B. In this situation the choice of the risky portfolio depends on the investor's coefficient of risk aversion, and there is an investor who is indifferent between these two portfolios. Specifically, the indifference between portfolios

A and B means that both portfolios provide the same utility. Formally, this condition yields

$$U(r_A) = U(r_B).$$

In particular,

$$\mu_A - \frac{1}{2}A\sigma_A^2 = \mu_B - \frac{1}{2}A\sigma_B^2.$$

With the solution

$$A^* = 2 \times \frac{\mu_A - \mu_B}{\sigma_A^2 - \sigma_B^2}.$$

That is, the investor with A^* is indifferent between risky portfolios A and B. In addition, we can easily deduce that more risk tolerant investors (who have $A < A^*$) prefer to choose portfolio A, whereas more risk averse investors (who have $A > A^*$) prefer to choose portfolio B.

Overall, in this section we demonstrated two important things. First, in the presence of the risk-free asset the Sharpe ratio facilitates the choice of the optimal risky portfolio.[30] However, without the solution of the optimal capital allocation problem the ultimate investor's goal, to maximize the utility of final wealth, is not achieved. Therefore if the investor is unable to select the optimal capital allocation, the use of the Sharpe ratio makes little or no sense. Second, in the absence of the risk-free asset the Sharpe ratio cannot be used at all; in this case the optimal risky portfolio is investor-specific.

9.7.2 The Asset Allocation Puzzles

Markowitz mean-variance portfolio theory, which is an important part of modern finance theory, is a sheer example of a normative theory. Specifically, Markowitz portfolio theory tells the investors how they ought to select optimal portfolios, but it does not explain how the investors select optimal portfolios in reality. In fact, the predictions of the mean-variance portfolio theory are in sharp contrast with the popular investment advice. This discrepancy between the theory and popular advice gives rise to the so-called "asset allocation puzzles", see Canner, Mankiw, and Weil (1997).

Consider the investor's allocation between cash (which serves as a risk-free asset), bonds, and stocks. Mean-variance portfolio theory predicts that all investors will select the same risky portfolio of stocks and bonds, the only difference will be in the capital allocation between the cash and the risky

---

[30] In this case the optimal risky portfolio is the same for all investors, see Chap. 7.

portfolio. More specifically, mean-variance portfolio theory predicts that the composition of the optimal risky portfolio of stocks and bonds will be the same for all investors regardless of their levels of risk aversion. In addition, the composition of the risky portfolio will be the same regardless of the investment horizon. This means that both short-term and long-term investors will select the same risky portfolio.

The popular investment advice from financial advisors is as follows. Financial advisors, first of all, divide all investors into several categories according to their willingness to take on risk (in other words, according to their risk aversion). For example, all investors can be divided into the following three broad categories: "conservative", "moderate", and "aggressive" (the names are self-explanatory). Then, for each type of investors, financial advisors recommend a specific composition of cash/bonds/stocks portfolio. For instance, conservative investors are advised to invest 40% in cash, 40% in bonds, and 20% in stocks. Aggressive investors, on the other hand, are advised to invest 5% in cash, 30% in bonds, and 65% in stocks. The first asset allocation puzzle, therefore, is that the investor's risk aversion influences the composition of his portfolio. Financial advisors also tend to recommend that the investor's time horizon should influence the composition of his portfolio; this gives rise to the second asset allocation puzzle. In particular, if the time horizon is long, investors should invest more aggressively. That is, if the investment horizon is long, more money should be allocated to stocks. As the investment horizon gets shorter, the weight of stocks in the portfolio should decrease, whereas the weight of bonds should increase.

Elton and Gruber (2000) show that relaxing the assumption about the existence of a risk-free asset allows one to explain the first asset allocation puzzle. Specifically, without a risk-free asset the composition of the investor's optimal portfolio depends on his risk aversion (see the previous section): more risk tolerant investors prefer to invest more in stocks, whereas more risk averse investors prefer to allocate more to bonds. To explain the second asset allocation puzzle is more challenging. It looks like that the only possible explanation of the second asset allocation puzzle is to assume that the investor's risk aversion depends on length of the investment horizon; yet this assumptions is not quite reasonable.

One of the serious weak points of modern finance theory in general, and Markowitz portfolio theory (as well as its equilibrium extension - the Capital Asset Pricing Model) in particular, is that these theories are built up on the assumption about the existence of a risk-free asset. This assumption significantly simplifies the selection of optimal portfolios and the construction of a market equilibrium model. This is because when a risk-free asset is present, the optimal portfolio is unique for all investors regardless of their risk preferences. Relaxing this assumption virtually destroys all existing capital market equilibrium models (including the models in the Arbitrage Pricing Theory). The other questionable assumption in modern finance theory is that risk can be adequately measured by standard deviation (that is, uncertainty). To emphasize the problem of using uncertainty as a risk measure, consider the following joke[31]:

> What is riskier - jumping out of an airplane with a parachute or jumping without one? The answer, surprisingly, depends on how you define risk. If your definition, like that of most investors, is the chance of a negative outcome - in this case, death - then without a parachute is the riskier. But if your definition of risk, like that of most finance professors, is uncertainty, then with a parachute is riskier: you may or may not die. If you jump without a parachute there is no uncertainty and, therefore, no risk.

In 2002 Daniel Kahneman received the Nobel Memorial Prize in Economics for the development of a behavioral finance theory (called Prospect theory) where the investors are loss averse (see Kahneman and Tversky, 1979). The idea of loss aversion is encapsulated in the expression "losses loom larger than gains" meaning that investors prefer avoiding losses to acquiring equivalent gains. That is, avoiding losses is the fundamental principle in making decisions under uncertainty. However, long before the advent of Prospect theory of Kahneman and Tversky, Benjamin Graham advocated for the "margin of safety" investment principle which is basically equivalent to the "avoiding losses" principle:

> Confronted with a challenge to distill the secret of sound investment into three words, we venture the motto, **Margin of Safety**. (Benjamin Graham, 1949, Chap. 16)

The term "margin of safety" was coined by Graham and Dodd already in their classical book "Security Analysis" from 1934. In this book, Graham proposed a clear definition of investment that was distinguished from what he deemed speculation:

> An investment operation is one which, upon thorough analysis promises safety of principal and an adequate return. Operations not meeting these requirements are speculative.

---

[31]This joke is found on http://www.theage.com.au/articles/2004/01/24/1074732659690.html.

Table 9.14 Probability of loss and mean return over different investment horizons for three major asset classes

Asset	Statistics	Investment horizon, years									
		1	2	3	4	5	6	7	8	9	10
Cash	Probability of loss, %	0	0	0	0	0	0	0	0	0	0
	Mean return, %	5	9	14	19	25	30	36	42	49	55
Bonds	Probability of loss, %	16	8	4	2	1	0	0	0	0	0
	Mean return, %	6	12	18	25	33	41	50	59	69	80
Stocks	Probability of loss, %	33	24	15	12	9	4	2	1	0	0
	Mean return, %	10	20	30	42	54	65	77	90	104	119

Graham carefully explains each of the key terms in his definition: "thorough analysis" means "the study of the facts in the light of established standards of safety and value" while "safety of principal" signifies "protection against loss under all normal or reasonably likely conditions or variations" and "adequate" (or "satisfactory") return refers to "any rate or amount of return, however low, which the investor is willing to accept, provided he acts with reasonable intelligence".

We conjecture that the popular investment advice is deeply rooted in Graham's investment philosophy which is, first and foremost, to preserve capital (termed as "safety of principal") and then to try to make it grow. It is worth to recap the two basic principles of Graham's investment philosophy:

1. The investor must deliberately protect himself against losses;
2. The investor must aspire to "adequate", not extraordinary, return.

To emphasize the differences between the three major asset classes (cash, bonds, and stocks), we use the monthly total return data on the S&P Composite index, the bond index, and the cash proxied by 1-month Treasury Bill rate. The data span the period from January 1926 to December 2011. The bond index return is an equally-weighted return on the long- and intermediate-term US government bonds; these data are provided by Ibbotson and Associates Inc.[32] We vary the investment horizon from 1 to 10 years, and for each asset class we compute the probability of loss and mean return. The probability of loss is the probability of a negative return over an investment horizon of specific length; this probability is the probability that the initial value of the principal will not be preserved by the end of a specific investment horizon. The mean return is the mean return over an investment horizon of specific length.

---

[32]More specifically, these data are from the Ibbotson SBBI 2012 Classic Yearbook.

Table 9.14 reports the results of estimating the probabilities of loss and mean returns for three major asset classes and different investment horizons. The data in this table allow us to explain the popular investment advice and the second asset allocation puzzle in the light of Graham's investment philosophy. The first observation is that, regardless of the length of investment horizon, the mean return to stocks is higher than the mean return to bonds which is higher than the mean return to cash. In other words, stocks are more rewarding than bonds that are more rewarding than cash. However, when it comes to the safety of principal, over short- to medium-term horizons cash is safer than bonds that are safer than stocks. It is worth noting that cash is a safe asset regardless of the length of the investment horizon. Specifically, the probability of losing money on cash investment is zero even though the rate of return is uncertain. In other words, when the investor allocates money to cash, his return is uncertain over horizons longer than 1 month. Therefore according to modern finance theory cash is a risky asset for investments beyond 1 month. On the other hand, if risk is measured by losses, not uncertainty, then cash is a risk-free asset regardless of the length of the investment horizon.

If, for example, the investor wants to invest for only one year, the only asset that guaranties the safety of principal is cash. As a result, the weights of cash/bonds/stocks in the investor's portfolio should be (100%,0%,0%). However, if the investor wants to invest for 6 years, both cash and bonds guarantee[33] the safety of principal, but bonds provide the highest mean return. Therefore in this case it makes sense to invest initially in bonds. The weights of cash/bonds/stocks in the investor's portfolio in this case can be (0%,100%,0%). Yet, as the investment horizon shortens, to reduce the probability of loss the investor should gradually decrease the weight of bonds in his portfolio and increase the weight of cash. Finally, if the investor wants to invest for 10 years, both cash, bonds, and stocks guarantee the safety of principal, but stocks provide the highest mean return. In this case the weights of cash/bonds/stocks in the investor's initial portfolio might be (0%,0%,100%). As the investment horizon decreases to 7–8 years, the investor needs to withdraw some money from stocks and invest in bonds. When the investment horizon becomes 4–5 years, the investor should probably withdraw all money from stocks and allocate between cash and bonds.

Many financial advisors, as well as Benjamin Graham, advocate of always investing in a portfolio of stocks and bonds. By doing this the investor benefits from the effect of diversification. Diversification is a term that can be summed up with the familiar phrase: "don't put all your eggs in one basket". Because the

---

[33]The usual disclaimer applies. Our estimations are based on using the past data, but the past is not a guarantee of the future.

correlation between stocks and bonds returns is usually low,[34] bonds typically counteract stock market losses during bear markets. Even though a portfolio of stocks and bonds has a reduced mean return as compared to that of stocks, the reduction of risk through the diversification effect exceeds by far the reduction of mean return.

9.7.3   The Benefits of the Moving Average Strategy

We remind the reader the ultimate question we are trying to answer in this chapter: The investor considers investing either in the S&P Composite index (currently this index is identical to the S&P 500 index) or in the moving average strategy that switches between the S&P Composite index and the risk-free asset depending on the identified trend direction. The investor wants to know whether the moving average strategy will outperform the passive investment in the S&P Composite index in the future. To answer this question, using the past data and the forward-testing methodology we evaluated the outperformance delivered by the moving average strategy and tested whether the outperformance is statistically significant. Since our tests for structural breaks in the long-run dynamics of the S&P Composite index revealed a major break around 1944, in evaluating the expected future outperformance of the moving average strategy we need to focus on the outperformance during the post-World War II period.

The results of our forward tests suggest that the outperformance produced by the moving average trading strategy tends to be positive over a long run. However, this outperformance is not statistically significant at conventional statistical levels. On average, our tests say that there is a 70% probability that the estimated long-run outperformance is a "true outperformance", but there is a 30% probability that the outperformance is a result of randomness. In other words, the chances that the moving average strategy underperforms the buy-and-hold strategy over a long run are rather high. Therefore the results of our tests are encouraging, but inconclusive according to the strong scientific standards.

However, even though in our tests we used the contemporary "state of the art" performance measurement methodology that is employed in the papers published in the leading financial journals, one has always to keep in mind that this performance measurement theory is based on a number of assumptions. The first problem the investor must realize is that even if all assumptions are met in practice, and even if the results of tests present statistically significant

---

[34] In contrast, the correlation between cash and bonds returns is usually very high. Therefore a portfolio of cash and bonds is not benefited from the diversification effect.

evidence that one strategy outperforms the other, this knowledge is of little value unless the investor knows how to allocate optimally his wealth between the risk-free asset and the risky portfolio. For example, if for some investor it is optimal to invest 100% in stocks, and this investor is told that the moving average strategy outperforms the passive stock investment, there is absolutely no guarantee that by investing 100% in the moving average strategy the investor increases his (expected) utility of terminal wealth. Modern finance theory only tells in this case that using the moving average strategy is better than using the passive stock index, but in order to benefit from this knowledge the investor must allocate optimally between the risk-free asset and the moving average strategy. In order to optimally allocate wealth between the risky portfolio and the risk-free asset, the investor needs to know his risk aversion coefficient. We doubt that there is even a single investor who knows the value of his coefficient A in the mean-variance utility function. Therefore, modern finance theory is basically oriented toward the needs of portfolio managers, not toward the needs of investors. The second problem in performance measurement is that it assumes the existence of a risk-free asset and the possibility of unlimited borrowing. These assumptions are usually not met in practice which means that there is no unique solution to the optimal portfolio choice problem.

If we admit that there is no risk-free asset[35] in real markets, then the mean-variance portfolio theory says that the choice of a risky portfolio depends on the investor's risk preferences. We found that the moving average strategy is both less risky and less rewarding than the corresponding buy-and-hold strategy. Therefore, even within the framework of modern finance theory, in the absence of a risk-free asset more risk tolerant investors prefer to allocate to stocks, whereas more risk averse investors tend to allocate to the moving average strategy. That is, in the absence of a risk-free asset the choice between the buy-and-hold strategy and the moving average strategy depends on the investor's risk preferences; the moving average strategy should be preferred if the investor risk aversion is relatively high. Summing up, in the mean-variance framework of modern finance theory when there is no risk-free asset, we can draw the conclusion that the moving average strategy is likely to appeal to risk-averse investors.

In addition, our tests revealed that the main advantage of the moving average trading strategy lies in its superior downside protection. To emphasize this feature of the moving average strategy, we do the following trick: we compound the monthly returns to the moving average strategy and the corresponding

---

[35] At this moment we use the standard definition of a risk-free asset in modern finance theory. Specifically, a risk-free asset is an asset that provides a deterministic return. That is, there is no uncertainty in the future rate of return on this asset.

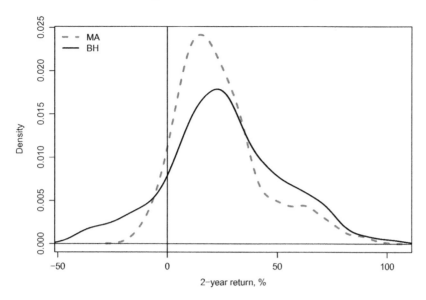

Fig. 9.16 Empirical probability distribution functions of 2-year returns on the buy-and-hold strategy and the moving average strategy. **BH** denotes the buy-and-hold strategy, whereas **MA** denotes the moving average trading strategy

buy-and-hold strategy to returns over 2-year periods.[36] Figure 9.16 plots the empirical probability distribution functions of 2-year returns on the buy-and-hold strategy and the moving average strategy.

Figure 9.16 advocates that the shapes of the two empirical probability distribution functions are rather different. Specifically, whereas the probability distribution function of 2-year returns to the buy-and-hold strategy has almost symmetrical shape around the mean, the probability distribution function of 2-year returns to the moving average strategy has a distinct right-skewed shape. It is important to observe that the empirical probability distribution functions differ mainly in the domain of losses, where the returns are negative. In contrast, in the domain of gains, where the returns are positive, the two distribution functions differ only a little. Since the probability of loss equals the area under the probability distribution function to the left of zero, we conclude that the probability of losing money over a 2-year horizon is much higher for the buy-and-hold strategy than for the moving average strategy.

The shapes of the two empirical probability distribution functions suggest that, when we compare the riskiness of the two alternative strategies using the

---

[36]Specifically, we use the period from January 1929 to December 1943 as the initial in-sample segment of data, and simulate the out-of-sample returns to the moving average strategy over January 1944 to December 2015 using an expanding in-sample window. The moving average strategy is based on selecting the best trading rule in a back test among 4 available rules: MOM, SMAC, SMAE, and EMACD. Then we compound the monthly out-of-sample returns to returns over 2-year periods.

standard deviation, we compare "apples and oranges". A correct comparison of riskiness requires taking into account the differences between the shapes of the two probability distribution functions. To provide a deeper insight into the comparative riskiness of several alternative strategies, in addition to the standard deviation we will also compute the skewness of the probability distribution and the probability of loss. Formally, the probability of loss is defined by

$$\text{Probability of loss} = Prob(r < 0),$$

where r denotes the return and $Prob(\cdot)$ denotes the probability. The problem in using the probability of loss as a risk measure is the fact that this measure tells nothing about the magnitude of potential loss if loss occurs. That is, in principle, one financial asset may have a higher probability of loss than the other asset, but the losses on the latter asset might be much more severe than the losses on the former asset. To complete the picture of losses, we will also compute the expected loss if loss occurs. This risk measure represents a specific realization of the popular risk measure that is known under different aliases: the Conditional Value-at-Risk (CVaR), the Expected Shortfall (ES), and the Expected Tail Loss (ETL). Formally, the expected loss if loss occurs is computed as

$$\text{Expected loss if loss occurs} = E[r|r < 0],$$

where $E[r|r < 0]$ denotes the expected return conditional on the outcome $r < 0$.

Besides the descriptive statistics of 2-year returns to stocks (that is, the buy-and-hold strategy) and the moving average strategy, we compute the descriptive statistics of 2-year returns to bonds, cash, and the 60/40 portfolio of stocks and bonds. As before, the bonds return is an equally-weighted return on the long- and intermediate-term US government bonds. The 60/40 portfolio of stocks and bonds is popular with pension funds and other long-term investors. This portfolio mix represents the "rule of thumb" for retirement portfolios. This portfolio mix also serves as a benchmark in most portfolio discussions.

Table 9.15 reports the descriptive statistics of 2-year returns for the three major asset classes as well as the descriptive statistics for the moving average strategy and the 60/40 portfolio mix. The assets in the table are ordered left-to-right by decreasing mean returns and standard deviation of returns. Observe that the moving average strategy is located in between the stocks and the 60/40 portfolio mix. This is because the moving average strategy has lower mean return than that of the stocks, but higher mean return than that of the 60/40 portfolio mix. Similarly, the moving average strategy has lower standard

Table 9.15   Descriptive statistics of 2-year returns on several alternative assets

Statistics	Stocks	MA	60/40	Bonds	Cash
Mean return, %	25.79	24.10	20.29	12.71	9.06
Standard deviation, %	26.43	20.04	16.81	12.35	6.20
Skewness	0.08	0.94	0.21	1.40	0.87
Probability of loss, %	13.75	7.06	10.47	9.84	0.00
Expected loss if loss occurs, %	−16.80	−5.20	−8.33	−2.16	

Notes MA denotes the moving average strategy whereas 60/40 denotes the 60/40 portfolio of stocks and bonds. The descriptive statistics are computed using the data over the period from January 1944 to December 2011

deviation of returns than that of the stocks, but higher standard deviation of return than that of the 60/40 portfolio mix. That is, judging by the mean-variance criterion, the moving average strategy is more rewarding than the 60/40 portfolio mix, but at the same time it is more risky.

On the other hand, when risk is measured by the probability of loss and expected loss, the moving average strategy proves to be less risky than the 60/40 portfolio mix. In other words, if we define risk as the chance of a negative outcome, then the moving average strategy is both more rewarding and less risky than the popular retirement portfolio. The fact, that the moving average strategy has higher standard deviation than that of the popular portfolio mix, appears mainly because the moving average strategy has higher variability (as compared with the 60/40 portfolio) in the domain of gains, which has nothing to do with riskiness. Interestingly, the moving average strategy has lower probability of loss than that of bonds. However, as revealed by the expected loss risk measure, losses on the moving average strategy tend to be more severe as compared with losses on bonds. Finally note that over a 2-year horizon the cash is clearly a risky asset if risk is measured by standard deviation. Specifically, the standard deviation of 2-year returns on cash amounts to about 6% which is about twice as low as that of bond returns. In contrast, when risk is measured by the probability of loss, the cash remains a risk-free asset.

Compared to the passive investment in stocks, the moving average strategy has a bit lower mean return, but at the same time substantially lower risk that is measured by the probability of loss and expected loss. In particular, the moving average strategy has twice (thrice) as low the probability of loss (the expected loss) as that of the buy-and-hold strategy.[37] Even though the

---

[37]This fact suggests that using the Sortino ratio for measuring the moving average strategy's outperformance makes much more sense than using the Sharpe ratio. However, replacing the Sharpe ratio with the Sortino ratio in out-of-sample tests does not influence the outcome of these tests: we cannot reject the hypothesis that the moving average strategy does not outperform the buy-and-hold strategy. This result agrees very well with the results reported by Zakamulin (2014) who also used the Sortino ratio for measuring the outperformance. One can logically assume that, for measuring the outperformance correctly, monthly

long-run growth from investing in stocks exceeds the long-run growth provided by the moving average strategy, over short- to medium-term horizons the moving average strategy is much less risky than the buy-and-hold strategy. Therefore the moving average strategy appeals not only to risk averse investors who invest for a long-run, but also to less risk averse investors who invest for a medium-run. As compared to the popular 60/40 portfolio mix, the moving average strategy seems to have a superior reward-to-risk combination. Thus, the moving average strategy seems to be a better retirement portfolio than the 60/40 portfolio. For the sake of illustration, Fig. 9.17 plots the cumulative returns to the moving average strategy versus the cumulative returns to the 60/40 portfolio of stocks and bonds.

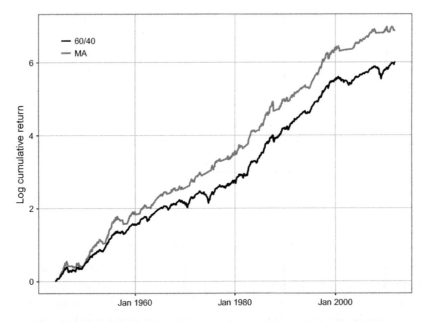

Fig. 9.17 Cumulative returns to the moving average strategy versus cumulative returns to the 60/40 portfolio of stocks and bonds over January 1944 to December 2011. MA denotes the moving average strategy whereas 60/40 denotes the 60/40 portfolio of stocks and bonds. The returns to the moving average strategy are simulated out-of-sample using an expanding in-sample window. The initial in-sample period is from January 1929 to December 1943

Lust but not least, the returns to the moving average strategy resemble the returns to a popular "portfolio insurance" strategy. In particular, traditional portfolio insurance strategy consists in investing in stocks and buying put

---

returns should be replaced by, for example, 2-year returns. However, in this case we have only 35 non-overlapping 2-year return observations during the out-of-sample period from 1944 to 2015. This sample is too small; the main issue with a small sample size is low power of statistical tests.

options on stocks as insurance. The price of these put options represents, in fact, the insurance premium the investor pays to buy portfolio insurance. During bull markets when stock prices trend upward, the insurance premium reduces the investor's return (because put options expire worthless). However, during bear markets when stock prices trend downward and the investor loses on the stocks, the portfolio insurance covers a part of the losses. As a result, the portfolio insurance strategy underperforms the buy-and-hold strategy during bull markets, but outperforms the buy-and-hold strategy during bear markets. Similarly, we found that the moving average strategy tends to underperform (outperform) the buy-and-hold strategy during bull (bear) markets. In contrast to the traditional portfolio insurance strategy, in the moving average strategy eventual losses on the stocks are "covered" only partially, and the amount of covered losses varies over time. Anyway, the moving average strategy represents a prudent investment strategy for a risk averse investor (or as a retirement portfolio) because its mean return and risk are reasonably consistent with his objectives and risk tolerance.

9.8   Chapter Summary

In this chapter we utilized the longest historical sample of monthly data on the S&P Composite stock market index with the goal to comprehensively evaluate the outperformance delivered by the moving average trading strategy. Yet while long history provides us with rich information about the past performance of moving average rules, the availability of long-term data is both a blessing and a curse. This is because in order to use the observed outperformance over a very long-term as a reliable estimate of the expected outperformance in the future, we need to make sure that the stock market dynamics both in the distant and near past were the same. However, the results from our robustness tests and tests for structural breaks revealed evidence of a major regime shift in the stock market dynamics that occurred around 1944. Specifically, starting from around 1944 the growth rate of the S&P Composite index has more than doubled. Most importantly, we found evidence that the average bull (bear) market duration has increased (decreased) over time. As compared with the first sub-period, over the second sub-period the ratio of the average bull market length to the average bear market length has almost doubled. Since the benefits of the moving average trading strategy come from timely identification of bear market states and moving to cash, it is only logical to conclude that the profitability of the moving average strategy has diminished over time.

We started our examination of the performance and properties of the moving average trading strategies by conducting back-tests. Even though the performance of the best trading rules in a back test is upward-biased and, therefore, it cannot be used as a reliable estimate of the expected future performance of these trading rules, the results of the back-tests allow us to draw the following useful conclusions:

- The short selling strategy, when the trader sells stocks short when a Sell signal is generated, is risky and does not pay off. Specifically, the performance of the short selling strategy is substantially worse than the performance of the corresponding strategy where the trader switches to cash. The poor performance of the short selling strategy is a result of the fact that the moving average strategy identifies the bull and bear stock market states with a poor precision.
- From a practical point of view, when either daily or monthly data are used, the choice of performance measure does not have a crucial influence on the selection of the best trading strategy in a back test. Therefore the Sharpe ratio, which has become the industry standard for measuring risk-adjusted performance, seems to be the most natural choice for performance measurement.
- From a practical point of view, the choice of moving average does not have a crucial influence on the performance of moving average trading strategies. In particular, regardless of the choice of moving average, the performance of the best trading strategy in a back test remains virtually intact. In this regard, the SMA can be preferred as the simplest, best known and best understood moving average.
- Using the monthly data, the best trading strategy in a back test over the post-1944 period is the SMAC(2,10) strategy. This trading strategy is also among the top 10 best trading strategies over the total historical sample. In particular, we found that the Moving Average Crossover rule, that uses one shorter SMA with window size of 2 months and one longer SMA with window size of 10 months, performs best in back tests.
- The price-change weighting function in the SMAC(2,10) strategy has a humped-shape form that differs only marginally from the decreasing shape of the price-change weighting function in the popular P-SMA(10) strategy.
- The SMAC(2,10) strategy identifies the direction of the stock price trend using the 10-month SMA. This allows us to estimate that the average lag time in identification of turning points in the stock price trend amounts to 4.5 months. Therefore, as a ballpark estimate, the duration of a bear market

should be at least 12 months in order to make the trend following strategy profitable.

- Even in a back test the performance of the SMAC(2,10) strategy is very uneven over time; this strategy might underperform the buy-and-hold strategy over relatively long periods. However, this finding should not be surprising because the moving average strategy is virtually doomed to underperform the buy-and-hold strategy during bull markets (when the moving average strategy generates some false Sell signals).
- The SMAC(2,10) strategy, which uses the averaging window of 10 months, is the optimal strategy over a long run that spans periods of many decades long. Over a period of one decade, the size of the optimal averaging window varies from 4 to 16 months. This result suggests that the SMAC(2,10) strategy is not a strategy that is optimal in any given historical period, but rather a strategy that is optimal on "average" over a long run.

In order to provide a reliable estimate of the real-life outperformance delivered by the moving average trading strategy, we performed forward (that is, out-of-sample) tests. Even though conventional wisdom says that the out-of-sample performance of a trading strategy provides an unbiased estimate of its real-life performance, we demonstrated a serious deficiency in the traditional out-of-sample testing procedure. Specifically, we demonstrated that the results of forward tests of profitability of moving average trading rules depend, sometimes crucially, on the choice of the historical period where the trading rules are tested and on the choice of split point between the initial in-sample and out-of-sample subsets. This is because both the in-sample and out-of-sample performance of the moving average trading strategy is very uneven over time. In this regard, our choices for historical periods and split points are made in order to provide the most objective, unbiased, and typical picture of the out-of-sample outperformance that is delivered by the moving average trading strategy. The results of our forward tests suggest the following conclusions:

- Over the out-of-sample period from 1870 to 2015 the moving average strategy tends to statistically significantly outperform the buy-and-hold strategy.
- Over the most relevant post-1944 out-of-sample period, our tests suggest that the moving average strategy tends to outperform the buy-and-hold strategy over a long run. However, this outperformance is not statistically significant at conventional statistical levels.
- Using daily data instead of monthly does not allow improving the out-of-sample performance of the moving average trading strategy. In fact, our results reveal that the out-of-sample performance of the moving average

strategy deteriorates when daily data are used instead of monthly. Our results also suggest that only the Moving Average Envelope (MAE) rule tends to outperform the buy-and-hold strategy in out-of-sample tests when daily data are used.

- Our results advocate that the out-of-sample performance of the moving average trading strategy tends to be better when an expanding in-sample window is used. The best out-of-sample performance is usually achieved when the in-sample window contains periods of severe market downturns.

- The moving average strategy has lower mean return and lower standard deviation of returns than those of the buy-and-hold strategy. The main advantage of the moving average trading strategy seems to be its superior downside protection. Specifically, the moving average strategy has substantially smaller drawdowns compared to the buy-and-hold strategy.

- The moving average strategy tends to underperform (outperform) the buy-and-hold strategy during bull (bear) markets. Even though the moving average strategy tends to outperform the buy-and-hold strategy during bear markets, this strategy also suffers losses during bear markets. However these losses are significantly smaller compared to losses suffered by the buy-and-hold strategy.

- The moving average strategy identifies the bull and bear states of the market with about 75% precision. In other words, the moving average strategy generates correct Buy and Sell signals about 75% of time. The estimated delay between the Bull-Bear states of the market and the periods of Buy-Sell trading signals amounts to 4 months. This number agrees very well with the average lag time of SMA(10) which amounts to 4.5 months.

- The out-of-sample outperformance is very uneven in time and is not guaranteed. In fact, our results suggest that over short- to medium-term horizons the market timing strategy is more likely to underperform the market than to outperform.

- The out-of-sample performance of trading rules that have a smaller number of parameters tends to be better than that of the rules that have a larger number of parameters. For instance, the out-of-sample performance of the P-SMA rule tends to be better than that of the SMAC rule. The performance of the P-SMA rule seems to be more robust than the performance of the MOM rule. This conclusion agrees with the results reported by Zakamulin (2015).

Armed with the results of numerous in-sample and out-of-sample tests, we are able now to revisit the myths regarding the superior performance of the moving average trading strategy. Specifically, many studies of the moving average trading strategy report that this strategy allows investors both to enhance returns and greatly reduce risk as compared to the buy-and-hold strategy. For example, Faber (2007) claims that the moving average trading strategy produces "equity-like returns with bond-like volatility and drawdowns". We can say with full confidence that this claim is obviously false. In particular, in out-of-sample tests the moving average strategy produces lower mean returns compared to the buy-and-hold strategy. As compared to bonds, the moving average strategy has higher volatility and larger drawdowns.

The other issue with many studies of the moving average trading strategy is that their results and claims create an illusion that the outperformance delivered by the moving average strategy is time invariant. The investors are deluded by a wrong belief that the moving average strategy always beats the buy-and-hold strategy. Such studies mislead the investors; many investors who invested in the moving average trading strategy from about 2009 have been utterly disappointed in the performance of this strategy because it underperformed the buy-and-hold strategy from 2009 to 2015 on year-to-year basis. These investors were not told that one has to expect that the moving average strategy underperforms during bull markets. Even during bear markets this strategy tends to underperform the buy-and-hold strategy when bear markets have a relatively short duration.

However, our results do not indicate that the market timing with moving averages has no sense. On the contrary, according to our evaluation the moving average trading strategy represents a prudent investment strategy for "moderate" and even "conservative" medium- and long-term investors. This is because the returns to the moving average trading strategy resemble the returns to the popular portfolio insurance strategy; the insurance premium reduces the returns if stock prices increase, but partially covers losses when stock prices decrease. As compared to the popular among long-term investors 60/40 portfolio of stocks and bonds, the moving average strategy seems to have a superior reward-to-risk combination. Specifically, when risk is measured by the probability of loss and the expected loss if loss occurs, the moving average strategy has both higher mean return and lower risk than the 60/40 portfolio mix.

Appendix 9.A: Testing for a Regime Shift in Stock Market Dynamics

The results reported in Table 9.1 suggest that the stock market mean (capital gain and total) returns and volatilities were different across the two sub-periods. To find out whether these differences are statistically significant, we perform the tests of the following null-hypotheses:

Equality of means: $H_0^1 : \mu_{1,CAP} = \mu_{2,CAP}$, $H_0^2 : \mu_{1,TOT} = \mu_{2,TOT}$,

Equality of variances: $H_0^3 : \sigma_{1,CAP}^2 = \sigma_{2,CAP}^2$, $H_0^4 : \sigma_{1,TOT}^2 = \sigma_{2,TOT}^2$,

where, for example, $\mu_{1,CAP}$ and $\mu_{2,CAP}$ denote the mean capital gain return during the first and the second sub-period respectively, and $\sigma_{1,CAP}$ and $\sigma_{2,CAP}$ denote the standard deviation of the capital gain return during the first and the second sub-period respectively. The first and the second null-hypotheses (H_0^1 and H_0^2) are standard null hypotheses for testing equality of two means. The third and the forth null-hypotheses (H_0^3 and H_0^4) are standard null hypotheses for testing equality of two variances. Since virtually all returns series exhibit non-normality and serial dependency, to test all the hypotheses we employ the stationary block-bootstrap method of Politis and Romano (1994).[38] Table 9.16 reports the results of the hypothesis tests. These results suggest that we have strong statistical evidence that the volatilities of both the capital gain and total returns have changed over time. We cannot reject the hypothesis that the mean total market return has been stable over time. However, at the 10% significance level we can reject the hypothesis about the stability of the mean capital gain returns over time.

Since our results advocate that there are economically and statistically significant differences in the mean capital gain returns across the two sub-periods of data, we perform an additional structural break analysis whose goal is twofold. The first goal is to verify that there is a major break in the growth rate of S&P Composite index. The second goal is to find the date of the breakpoint.

Table 9.16 Results of the hypothesis tests on the stability of means and standard deviations over two sub-periods of data

Hypothesis	p-value	Hypothesis	p-value
$H_0^1 : \mu_{1,CAP} = \mu_{2,CAP}$	0.09	$H_0^3 : \sigma_{1,CAP}^2 = \sigma_{2,CAP}^2$	0.00
$H_0^2 : \mu_{1,TOT} = \mu_{2,TOT}$	0.34	$H_0^4 : \sigma_{1,TOT}^2 = \sigma_{2,TOT}^2$	0.00

---

[38] For the description of the stationary bootstrap method, see Chap. 7.

Our null hypothesis is that the period t log capital gain return on the S&P Composite index, r_t, is normally distributed with constant mean μ and variance σ^2. More formally, $r_t \sim \mathcal{N}(\mu, \sigma^2)$. Under this hypothesis the log of the S&P Composite index at time t is given by the following linear model

$$\log(I_t) = \log(I_0) + \sum_{i=1}^{t} r_i = \log(I_0) + \mu t + \varepsilon_t, \qquad (9.2)$$

where I_0 is the index value at time 0 and $\varepsilon_t \sim \mathcal{N}(0, \sigma^2 t)$. Our alternative hypothesis is that the mean log capital gain return on the S&P Composite index varies over time. To test the null hypothesis, there are many formal tests (see Zeileis et al. 2003, and references therein). Unfortunately, the error term in regression (9.2) does not satisfy the standard i.i.d. assumptions (because ε_t exhibits heteroskedasticity and autocorrelation) and therefore these tests are not applicable in our case.

Our simplified alternative hypothesis is that the mean log return at time t^* changes from μ to $\mu + \delta$. Under the alternative hypothesis the log of the S&P Composite index at time t is given by the following segmented model

$$\log(I_t) = \log(I_0) + \mu t + \delta(t - t^*)^+ + \varepsilon_t, \qquad (9.3)$$

where $(t - t^*)^+$ denotes the positive part of the difference $(t - t^*)$. In this case the natural test of the null hypothesis is

$$H_0 : \delta = 0.$$

We find the breakpoint t^* using the methodology presented in Muggeo (2003). Both the models (given by equations (9.2) and (9.3)) are estimated using the total sample period 1857–2015. The results of the estimations of the two alternative models are reported in Table 9.17. The p-values of the estimated coefficients are computed using the heteroskedasticity and autocorrelation consistent standard errors.

Apparently, the we can reject the null hypothesis of constant log mean return at the 1% significance level. The segmented model has a higher R-squared (98% versus 90% for the linear model) and double as low the residual standard deviation (27% versus 62% for the linear model). The estimated date of the breakpoint is September 1944; therefore January 1944 is chosen as the start of the second sub-period of our data. The 95% confidence interval for the breakpoint date is from September 1943 to September 1945. Under the assumption of constant mean log returns, over the total sample period

Table 9.17 Results of the estimations of the two alternative models using the total sample period 1857–2015

	Linear model	Segmented model
Intercept $\log(I_0)$	−7.05e-01	1.62e-01
	(0.00)	(0.00)
Coefficient μ	3.45e-03	1.73e-03
	(0.00)	(0.00)
Coefficient δ		4.11e-03
		(0.00)
Adjusted R-squared	0.90	0.98
Residual std. deviation	0.62	0.27

Notes The linear model is given by $\log(I_t) = \log(I_0) + \mu t + \varepsilon_t$. The segmented model is given by $\log(I_t) = \log(I_0) + \mu t + \delta(t - t^*)^+ + \varepsilon_t$. The p-values of the estimated coefficients are given in brackets. The estimated breakpoint date is September 1944

(that spans 159 years) the estimated mean log return amounts to approximately 4% in annualized terms. However, this assumption proofs to be wrong and a more detailed examination of the growth rate of the log of the S&P Composite index suggests that around year 1944 (87 years from the start of the sample) there was a major break in the growth rate. Specifically, prior to 1944 the estimated mean log return was about 2%, thereafter about 7% in annualized terms.

Appendix 9.B: Testing for a Structural Break in Bull-Bear Dynamics

Consider a two-state Markov switching model for returns where S_t denotes the latent state variable at time t. The state variable can take one of two possible values: 0 (denotes a Bear market state) and 1 (denotes a Bull market state). This Markov switching model for returns in sub-period $m \in \{1, 2\}$ can be written as

$$r_t^m | S_t \sim N\left(\mu_{S_t}^m, \left(\sigma_{S_t}^m\right)^2\right),$$
$$p_{ij}^m = P^m(S_t = j | S_{t-1} = i),$$

where $i, j \in \{0, 1\}$. This model assumes that the stock market returns at time t of sub-period m are normally distributed with mean μ_0^m and standard deviation σ_0^m if the market is in state 0. Otherwise, in state 1, the stock market returns are normally distributed with mean μ_1^m and standard deviation σ_1^m. p_{ij}^m is the probability of transition from state i to state j in sub-period m. The transition probability matrix in sub-period m is given by

$$P^m = \begin{bmatrix} p_{00}^m & p_{01}^m \\ p_{10}^m & p_{11}^m \end{bmatrix}.$$

For example, p_{00} is the transition probability from a bear market to a bear market (or the probability that the market remains in Bear state), while $p_{01} = 1 - p_{00}$ is the transition probability from a bear market to a bull market. If at time t the market is in the Bear state, then at the next time $t+1$ the market remains in the Bear state with probability p_{00} or transits to the Bull state with probability p_{01}. Observe that the lower the transition probability p_{01}, the longer the market remains in the Bear state and, consequently, the longer the average duration of bear markets.

To find out whether the parameters of the bull and bear markets are the same in both sub-periods, we test the following set of null-hypotheses:

Equality of means: $H_0^1 : \mu_0^1 = \mu_0^2, \quad H_0^2 : \mu_1^1 = \mu_1^2,$

Equality of variances: $H_0^3 : \left(\sigma_0^1\right)^2 = \left(\sigma_0^2\right)^2, \ H_0^4 : \left(\sigma_1^1\right)^2 = \left(\sigma_1^2\right)^2,$

Equality of probability transition matrices: $H_0^5 : P^1 = P^2.$

To test hypotheses 1–2, we perform a standard two-sample t-test for equal means. To test hypotheses 3–4, we perform a standard two-sample F-test for equal variances.

We test the equalities of the two transition probability matrices by performing element-by-element tests of the stability of each entry p_{ij}^m. To estimate the transition probability p_{ij}^m and standard errors of estimation of p_{ij}^m, we use a bootstrap estimation approach proposed by Kulperger and Rao (1989). The bootstrap approach follows these steps: First, using the original data sequence of Bull and Bear markets, we estimate the transition probability matrix by employing the maximum likelihood estimator. Second, we generate 100 bootstrap samples of the data sequences following the conditional distributions of states estimated from the original one. Third, we apply maximum likelihood estimation on each bootstrapped data sequence. Forth, the estimated transition probability is computed as the average of all maximum likelihood estimators. Finally, after computing the average, we compute the standard deviation of our estimator and corresponding standard error of estimation. The hypothesis

Table 9.18 Estimated transition probabilities of the two-states Markov switching model for the stock market returns over two historical sub-periods: 1857–1943 and 1944–2015

	1857–1943		1943–2015	
	Bear	Bull	Bear	Bull
Bear	0.939	0.061	0.916	0.084
Bull	0.042	0.958	0.030	0.970

$H_0^{5_q} : p_{ij}^1 = p_{ij}^2, q \in \{1, 2, 3, 4\}$, is tested assuming that errors are normally distributed.

Table 9.18 reports the estimated transition probabilities of the Markov switching model for the stock market states over the two historical sub-periods. The comparison of the values of transition probabilities over the two historical sub-periods also advocates that the duration of the bear (bull) markets has decreased (increased) over time. Specifically, $p_{01}^1 = 0.061$ whereas $p_{01}^2 = 0.084$. This says that during the first sub-period the transition probability from a bear to a bull market was 6.1%, whereas over the second sub-period the transition probability from a bear to a bull market was 8.4%. That is, the transition probability from a Bear state to a Bull state has increased over time. As a consequence, the average length of bear markets has become shorter over time. Similarly, $p_{10}^1 = 0.042$ whereas $p_{10}^2 = 0.030$. This says that during the first sub-period the transition probability from a bull to a bear market was 4.1%, whereas over the second sub-period the transition probability from a bull to a bear market was 3.0%. As a result, the average duration of bull markets has become longer over time.

Table 9.19 reports the results of the hypothesis tests. These results suggest that we have strong statistical evidence that all the transition probabilities between the states of the stock market have changed over time (that is, we can reject the equality of the transition probability matrices over the two sub-periods), the mean stock market return during Bull states has changed over time, and that the volatility of the states have changed over time as well. Yet, we cannot reject the hypothesis that the mean stock market return during Bear states has been stable over time.

Table 9.19 Results of the hypothesis testing on the stability of the parameters of the two-states Markov switching model for the stock market returns over the two sub-periods

Hypothesis	p-value
$H_0^1 : \mu_0^1 = \mu_0^2$	0.93
$H_0^2 : \mu_1^1 = \mu_1^2$	0.04
$H_0^3 : (\sigma_0^1)^2 = (\sigma_0^2)^2$	0.00
$H_0^4 : (\sigma_1^1)^2 = (\sigma_1^2)^2$	0.00
$H_0^{51} : p_{00}^1 = p_{00}^2$	0.00
$H_0^{52} : p_{01}^1 = p_{01}^2$	0.00
$H_0^{53} : p_{10}^1 = p_{10}^2$	0.00
$H_0^{54} : p_{11}^1 = p_{11}^2$	0.00

References

Appel, G. (2005). *Technical analysis: Power tools for active investors.* FT Prentice Hall.

Auer, B. R. (2015). Does the choice of performance measure influence the evaluation of commodity investments? *International Review of Financial Analysis, 38,* 142–150.

Berk, J., & DeMarzo, P. (2013). *Corporate finance.* Pearson.

Bry, G., & Boschan, C. (1971). *Cyclical analysis of time series: Selected procedures and computer programs.* NBER.

Campbell, J. Y., & Shiller, R. J. (1998). Valuation ratios and the long-run stock market outlook. *Journal of Portfolio Management, 24*(2), 11–26.

Canner, N., Mankiw, N., & Weil, D. (1997). An asset allocation puzzle. *American Economic Review, 87*(1), 181–191.

Cherny, A., & Madan, D. (2009). New measures for performance evaluation. *Review of Financial Studies, 22*(7), 2571–2606.

Cogneau, P., & Hübner, G. (2009). The (More Than) 100 ways to measure portfolio performance. *Journal of Performance Measurement, 13,* 56–71.

Clare, A., Seaton, J., Smith, P. N., & Thomas, S. (2013). Breaking into the blackbox: Trend following, stop losses and the frequency of trading: The case of the S&P500. *Journal of Asset Management, 14*(3), 182–194.

Eling, M. (2008). Does the measure matter in the mutual fund industry? *Financial Analysts Journal, 64*(3), 54–66.

Eling, M., & Schuhmacher, F. (2007). Does the choice of performance measure influence the evaluation of hedge funds? *Journal of Banking and Finance, 31*(9), 2632–2647.

Elton, E., & Gruber, M. (2000). The rationality of asset allocation recommendations. *Journal of Financial and Quantitative Analysis, 35*(1), 27–41.

Faber, M. T. (2007). A quantitative approach to tactical asset allocation. *Journal of Wealth Management, 9*(4), 69–79.

Gartley, H. M. (1935). *Profits in the stock market.* Lambert Gann Publications.

Gonzalez, L., Powell, J. G., Shi, J., & Wilson, A. (2005). Two centuries of bull and bear market cycles. *International Review of Economics and Finance, 14*(4), 469–486.

Graham, B. (1949). *The intelligent investor*. New York: Harper and Brothers.

Graham, B., & Dodd, D. (1934). *Security analysis*. New York: Whittlesey House.

Kahneman, D., & Tversky, A. (1979). Prospect theory: An analysis of decision under risk. *Econometrica, 47*(2), 263–291.

Kulperger, R. J., & Rao, B. L. S. P. (1989). Bootstrapping a finite state markov chain. *Sankhya: Indian Journal of Statistics, Series A, 51*(2), 178–191.

Lunde, A., & Timmermann, A. (2004). Duration dependence in stock prices: An analysis of bull and bear markets. *Journal of Business and Economic Statistics, 22*(3), 253–273.

Muggeo, V. M. R. (2003). Estimating regression models with unknown break-points. *Statistics in Medicine, 22*(19), 3055–3071.

Pagan, A. R., & Sossounov, K. A. (2003). A simple framework for analysing bull and bear markets. *Journal of Applied Econometrics, 18*(1), 23–46.

Politis, D., & Romano, J. (1994). The stationary bootstrap. *Journal of the American Statistical Association, 89*, 1303–1313.

Schwert, G. W. (1990). Indexes of United States stock prices from 1802 to 1987. *Journal of Business, 63*(3), 399–442.

Shiller, R. J. (1989). *Market volatility*. The MIT Press.

Shiller, R. J. (2000). *Irrational exuberance*. Princeton University Press.

Walsh, C. E. (1993). Federal Reserve Independence and the Accord of 1951, FRBSF Economic Letter, 21. http://www.bus.lsu.edu/mcmillin/personal/4560/accord.html, [Online; Accessed 3-February-2017].

Welch, I., & Goyal, A. (2008). A comprehensive look at the empirical performance of equity premium prediction. *Review of Financial Studies, 21*(4), 1455–1508.

Zakamulin, V. (2010). On the consistent use of VaR in portfolio performance evaluation: A cautionary note. *Journal of Portfolio Management, 37*(1), 92–104.

Zakamulin, V. (2014). The real-life performance of market timing with moving average and time-series momentum rules. *Journal of Asset Management, 15*(4), 261–278.

Zakamulin, V. (2015). *Market timing with a robust moving average* (Working paper, University of Agder, https://papers.ssrn.com/sol3/papers.cfm?abstract_id=2612307)

Zeileis, A., Kleiber, C., Krämer, W., & Hornik, K. (2003). Testing and dating of structural changes in practice. *Computational Statistics and Data Analysis, 44*(1–2), 109–123.

10

Trading in Other Financial Markets

10.1 The Set of Tested Strategies and General Methodology

The majority of our data come at the monthly frequency; for some markets we have corresponding data at the daily frequency. Using monthly data, the following set of rules are back-tested:

MOM(n) for $n \in [2, 25]$, totally 24 trading strategies;
SMAC(s, l) for $s \in [1, 12]$ and $l \in [2, 25]$, totally 222 trading strategies;
SMAE(n, p) for $n \in [2, 25]$ and $p \in [0.25, 0.5, \ldots, 10.0]$, totally 1,060 trading strategies;

Using daily data, the following set of rules are back-tested:

MOM(n) for $n \in [2, 3, \ldots, 15, 20, 30, \ldots, 350]$, totally 48 trading strategies;
SMAC(s, l) for $s \in [1, 2, \ldots, 20, 25, \ldots, 80]$ and $l \in [2, 3, \ldots, 15, 20, 30, \ldots, 350]$, totally 1,144 trading strategies;
SMAE(n, p) for $n \in [2, 3, \ldots, 15, 20, 30, \ldots, 350]$ and $p \in [0.25, 0.5, \ldots, 10.0]$, totally 1920 trading strategies;

With monthly (daily) data, for each financial asset the overall number of tested trading strategies amounts to 1,206 (3,112).

We do not include the MACD rule in the set of tested rules. Because this rule is very flexible and easier to fit to data than the other rules, this rule tends to be over-represented among the best trading rules in a back test. However,

© The Author(s) 2017
V. Zakamulin, *Market Timing with Moving Averages*, New Developments
in Quantitative Trading and Investment, DOI 10.1007/978-3-319-60970-6_10

223

because this rule is prone to overfit the data, this rule usually delivers a poor performance in forward tests.

Regardless of the data frequency, in stock markets the returns to all strategies are simulated accounting for 0.25% one-way transaction costs; in all other markets the returns to all strategies are simulated accounting for 0.1% one-way transaction costs. In all strategies a Sell signal is usually a signal to leave the market and move to cash (or stay invested in cash). In currency and commodity markets we also investigate the performance of the strategy with short sales. The performance of all strategies is measured using the Sharpe ratio.

In our forward tests, the set of tested trading rules is the same as in back tests with one extension. Specifically, we add the P-SMA(n) rule where $n \in [2, 25]$ with monthly trading and $n \in [2, 3, \ldots, 15, 20, 30, \ldots, 350]$ with daily trading. Therefore with monthly (daily) trading, the total number of tested strategies amounts to 1,084 (3,160). The forward tests are implemented with an expanding in-sample window. With monthly trading the selection of the best trading strategy in the in-sample window is repeated every month. With daily trading, to speed up the simulation of the out-of-sample strategy, the selection of the best trading strategy in the in-sample window is repeated every 21th day. The null hypothesis of no outperformance is tested using the stationary block-bootstrap method consisting in drawing 10,000 random resamples with the average block length of 5 months.

10.2 Stock Markets

10.2.1 Data

In this section we use monthly and daily data on five stock market indices in the US (as well as the data on the risk-free rate of return). They are the Dow Jones Industrial Average (DJIA) index, the large cap stock index, the small cap stock index, the growth stock index, and the value stock index. All data span the period from July 1926 to December 2015. Until the end of 1952, stock exchanges in the US were open 6 days a week. Beginning from 1953, stocks were traded 5 days a week only. Therefore, for the sake of consistency of daily data series, we remove the return observations on Saturdays; the return on each removed Saturday is added to the return on the next trading day.

The DJIA index is a price-weighted stock index. Specifically, the DJIA is an index of the prices of 30 large US corporations selected to represent a cross-section of US industry. The components of the DJIA have changed 51 times in its 120 year history. Changes in the composition of the DJIA are made to reflect changes in the companies and in the economy. The daily DJIA index values for the total sample period and dividends for the period 1988 to 2015 are provided by S&P Dow Jones Indices LLC, a subsidiary of the McGraw-Hill Companies.[1] The dividends for the period 1926 to 1987 are obtained from Barron's.[2] Dividends are 12-month moving sums of dividends paid on the DJIA index. The monthly data series are obtained from daily data series using the close index values at the end of each calendar month. Daily and monthly index values are used to compute the daily and monthly capital gain returns respectively. The daily dividend yield is the simple daily yield that, over the number of trading days in the month, compounds to 1-month dividend yield. The total returns are obtained by summing up the capital gain returns and the dividend yields.

All other data are obtained from the data library of Kenneth French.[3] The returns on the large (small) cap index are the returns on the value-weighted portfolio consisting of the top (bottom) quintile (20%) of all of the firms in the aggregate US stock market after these firms have been sorted by their market capitalization. The number of stocks in the large cap index varies from 100 to 500. Thus, the return on the large cap index roughly corresponds to the return on the S&P Composite stock price index. Therefore the results for the large cap index can be used to check the robustness of our results for the S&P Composite index; we expect that the moving average strategy delivers similar outperformance for both the large cap index and the S&P Composite index. The returns on the growth (value) stock index are the returns on the value-weighted portfolio consisting of the top (bottom) quintile of all of the firms in the aggregate US stock market after these firms have been sorted by their book-to-market ratios.

All monthly data contain both the capital gain returns and total returns. When monthly data are used, trading signals are computed using the prices not adjusted for dividends. However, all daily data, but the data for the DJIA, contain the total returns only; the daily data on capital gain returns are not available in the data library of Kenneth French. Therefore, when daily data are used, trading signals are computed using the prices adjusted for dividends.

By definition, growth stocks (a.k.a. the "glamour" stocks) are stocks of companies that generate substantial cash flow and whose earnings are expected to grow at a faster rate than that of an average company. Value stocks are stocks that tend to trade at a lower price relative to its fundamentals and thus considered undervalued by investors. Common charac-

---

[1] http://www.djaverages.com.

[2] http://online.barrons.com.

[3] http://mba.tuck.dartmouth.edu/pages/faculty/ken.french/data_library.html.

teristics of such stocks include a high dividend yield, low price-to-book ratio, and low price-to-earnings ratio. Small cap stocks are stocks of companies with a relatively small market capitalization. Both the small stocks and value stocks are riskier than the large stocks, but at the same time they are more rewarding. Historically, the small stocks and value stocks outperformed the stock market as a whole (as well as the large stocks) on the risk-adjusted bias judging by either the Sharpe ratio or the alpha in the Capital Asset Pricing Model. The growth stocks, on the other hand, are a bit more risky than the large stocks and, at the same time, underperform a little the large stocks.

10.2.2   Back-Testing Trading Rules

Table 10.1 shows the top 10 best trading strategies in a back test over the period from January 1944 to December 2015. The results reported in this table suggest the following observations:

- With monthly trading, the SMAE rule is over-represented among the top 10 best trading strategies. With daily trading, virtually all top 10 best trading strategies are based on using the SMAE rule. This result advocates that the SMAE rule is superior to both the MOM and SMAC rules.
- With monthly trading, the SMAC(2,10) strategy is the best strategy for trading the S&P Composite index (in a back test over 1944–2015). The SMAC(2,10) strategy is also the second best strategy in trading the growth stocks. For the large stocks, the best trading strategy is the SMAC(2,11) strategy; the SMAC(2,12) is also among the top 10 best trading strategies.
- Outperformance delivered by the moving average trading rules depends on the stock index. Specifically, outperformance is the largest for the small stocks and the lowest for the DJIA index.
- In a back test, trading daily versus monthly allows the trader to improve the outperformance. The advantage in trading daily is the lowest for the DJIA index and the largest for the small stocks. In particular, for the small stocks the outperformance with daily trading is triple as much as the outperformance with monthly trading.
- The optimal size of the averaging window depends on the stock index. For trading the DJIA, the large stocks, and the growth stocks, the optimal size of the averaging window varies in between 190 and 220 days. For trading the value stocks, the optimal size of the averaging window varies in between 100 and 120 days. Finally, for trading the small stocks the optimal size of the averaging window varies in between 15 and 30 days.

Table 10.1 Top 10 best trading strategies in a back test

Rank	DJIA Strategy	Δ	Large stocks Strategy	Δ	Small stocks Strategy	Δ	Growth stocks Strategy	Δ	Value stocks Strategy	Δ
Trading at the monthly frequency										
1	SMAE(5,5)	0.06	SMAC(2,11)	0.13	SMAE(2,0.25)	0.31	SMAE(10,1)	0.07	SMAE(3,2.75)	0.10
2	SMAE(6,4.75)	0.06	SMAE(9,2)	0.12	MOM(2)	0.31	SMAC(2,10)	0.07	SMAE(4,3)	0.08
3	SMAE(6,5)	0.04	SMAE(10,2.25)	0.12	P-SMA(2)	0.31	SMAC(2,11)	0.07	SMAE(4,2.75)	0.08
4	SMAE(8,3)	0.03	SMAE(7,3.25)	0.12	SMAE(2,0.5)	0.20	SMAE(9,1.75)	0.07	SMAE(10,1.75)	0.07
5	SMAC(2,8)	0.03	SMAE(15,0.75)	0.12	SMAE(2,0.75)	0.18	SMAE(13,4.5)	0.07	SMAE(3,3)	0.07
6	SMAC(3,9)	0.03	SMAE(8,1.75)	0.12	P-SMA(4)	0.17	SMAE(12,1.5)	0.07	SMAE(3,3.25)	0.07
7	SMAE(7,5)	0.03	SMAE(11,1)	0.11	P-SMA(3)	0.17	SMAE(12,0.25)	0.06	SMAE(4,2.5)	0.06
8	SMAC(3,8)	0.02	SMAC(2,12)	0.11	SMAE(3,0.75)	0.17	SMAE(15,1.5)	0.06	SMAC(3,9)	0.06
9	SMAE(8,3.25)	0.02	SMAE(6,2.25)	0.11	SMAE(3,0.5)	0.17	SMAE(11,0.5)	0.06	MOM(6)	0.06
10	SMAE(6,4.25)	0.02	SMAE(7,2)	0.11	SMAE(3,1.25)	0.16	SMAE(14,3.25)	0.06	SMAE(3,2.5)	0.06
Trading at the daily frequency										
1	SMAE(220,3.25)	0.08	SMAE(220,2)	0.21	SMAE(30,0.5)	1.00	SMAE(160,3.5)	0.17	SMAE(110,0.5)	0.24
2	SMAE(200,3.75)	0.08	SMAE(210,2.25)	0.21	SMAE(30,1)	0.99	SMAE(220,3.5)	0.16	SMAE(110,0.25)	0.23
3	SMAE(210,3.5)	0.07	SMAE(220,3)	0.20	SMAE(30,0.25)	0.98	SMAE(220,3.75)	0.16	SMAE(120,0.5)	0.23
4	SMAE(220,3.5)	0.07	SMAE(200,2.75)	0.20	SMAE(30,1.25)	0.96	SMAE(250,1.75)	0.16	SMAE(120,0.75)	0.22
5	SMAE(190,3.75)	0.07	SMAE(210,2.5)	0.20	SMAE(12,1)	0.96	SMAE(230,3.75)	0.15	SMAE(110,0.75)	0.21
6	SMAE(190,4)	0.07	SMAE(200,2.5)	0.20	SMAE(20,0.25)	0.96	SMAE(150,4)	0.15	SMAE(100,0.25)	0.21
7	SMAE(210,3.25)	0.07	SMAE(190,2.25)	0.20	SMAE(15,1)	0.96	SMAE(150,4.5)	0.15	P-SMA(110)	0.21
8	SMAE(210,3.75)	0.07	SMAE(210,3)	0.20	SMAE(20,0.5)	0.95	SMAE(220,4)	0.15	SMAE(110,1)	0.20
9	SMAE(160,2.75)	0.07	SMAE(210,2)	0.20	SMAE(20,1)	0.94	SMAE(210,4.25)	0.15	SMAC(3,100)	0.20
10	SMAE(200,4)	0.06	SMAE(150,3.5)	0.20	SMAE(13,1)	0.94	SMAE(230,3.5)	0.15	SMAE(120,1)	0.20

Notes $\Delta = SR_{MA} - SR_{BH}$ where SR_{MA} and SR_{BH} are the Sharpe ratios of the moving average strategy and the buy-and-hold strategy respectively. The historical period is from January 1944 to December 2015

The reader is reminded that in our tests we always take into account transaction costs. The results on the best trading strategy in a back test in the absence of transaction costs are completely different. In particular, with daily trading without transaction costs, for virtually all stock market indices the MOM(2) strategy is the best trading strategy in a back test. Note that this strategy consists in buying the stocks when the daily price change is positive, and selling the stocks otherwise; this strategy exploits a very short-term momentum. For all stock market indices but the DJIA index, Fig. 10.1 plots the rolling performance (in the absence of transaction costs) of the MOM(2) strategy over the period from January 1927 to December 2015. The graphs in this plot suggest that the outperformance of this strategy was positive and increasing over the period from the early 1940s to the early 1970s. Afterwards, the outperformance delivered by the MOM(2) strategy was decreasing. This very short-

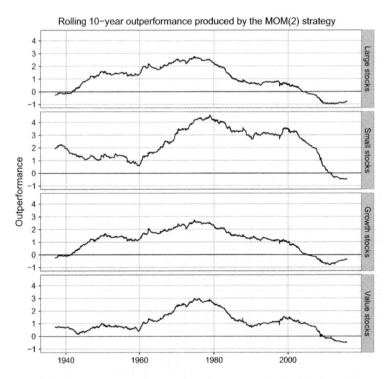

Fig. 10.1 Rolling 10-year outperformance produced by the MOM(2) strategy over the period from January 1927 to December 2015. The first point in the graph gives the outperformance over the first 10-year period from January 1927 to December 1936. The returns to the MOM(2) strategy are simulated assuming daily trading without transaction costs. Outperformance is measured by $\Delta = SR_{MA} - SR_{BH}$ where SR_{MA} and SR_{BH} are the Sharpe ratios of the moving average strategy and the buy-and-hold strategy respectively

term momentum in stock prices was especially strong in small stocks over the period from the early 1960s to the early 2000s. From about the mid-2000s the MOM(2) strategy started to underperform the buy-and-hold strategy. This fact suggests that the very short-term momentum in daily stock prices ceased to exist and was replaced by a very short-term mean-reversion.

10.2.3 Forward-Testing Trading Rules

The initial in-sample period is from January 1929 to December 1943. Consequently, the out-of-sample period is from January 1944 to December 2015. Table 10.2 reports the outperformance delivered by the moving average trading strategies in out-of-sample tests with monthly and daily trading. Our first observation is that regardless of the data frequency the moving average trading rules underperform the buy-and-hold strategy in trading the DJIA index and the growth stock index. This finding suggests that the moving average rules do not work in some stock markets. Our second observation is that the moving average trading rules statistically significantly outperform the buy-and-hold strategy in trading the small stocks. Daily trading the small stocks produces a much greater outperformance than monthly trading. Specifically, with daily trading the outperformance is from 4 to 7 times higher than that with monthly trading. Our third observation is that in trading the large stocks the outperformance is positive but is not statistically significant. Daily trading the large stocks has no advantages compared with monthly trading. These results for trading the large stocks agree very well with the results for trading the S&P Composite index. Our last observation is that in trading the value stocks some rules deliver a positive outperformance, but this outperformance is not statistically significant. The results for these stocks seem to suggest that daily trading has a small advantage compared with monthly trading.

Comparing the results of the forward tests with those of the back tests, we can note some similarities. Specifically, in trading the DJIA index the outperformance delivered by the best trading rules in a back test is marginal; in a forward test the moving average rules underperform the buy-and-hold strategy. In back tests, daily trading the small stocks produces significantly higher outperformance than monthly trading. Similarly, in forward tests, daily trading the small stocks produces significantly higher outperformance than monthly trading. Apparently, daily trading the small stocks was advantageous because the moving average rules exploited a strong short-term momentum existed in this market.

To gain further insights into the properties of the moving average trading strategy, we analyze the out-of-sample performance of the combined moving

Table 10.2 Outperformance delivered by the moving average trading strategies in out-of-sample tests

Stock index	Statistics	Moving average strategy				
		MOM	P-SMA	SMAC	SMAE	COMBI
Trading at the monthly frequency						
DJIA	Outperformance	−0.15	−0.05	−0.09	−0.08	−0.07
	P-value	0.96	0.73	0.84	0.84	0.80
Large stocks	Outperformance	0.04	0.07	0.02	0.06	0.01
	P-value	0.37	0.23	0.44	0.26	0.47
Small stocks	Outperformance	**0.11**	**0.20**	**0.12**	**0.11**	**0.10**
	P-value	0.10	0.01	0.10	0.09	0.10
Growth stocks	Outperformance	−0.04	−0.00	−0.01	−0.02	−0.01
	P-value	0.68	0.53	0.54	0.61	0.58
Value stocks	Outperformance	−0.12	−0.08	−0.04	0.03	0.03
	P-value	0.91	0.80	0.69	0.39	0.38
Trading at the daily frequency						
DJIA	Outperformance	−0.28	−0.05	−0.05	−0.03	−0.03
	P-value	1.00	0.71	0.71	0.62	0.62
Large stocks	Outperformance	−0.01	0.04	0.02	0.05	0.05
	P-value	0.53	0.33	0.41	0.30	0.30
Small stocks	Outperformance	**0.74**	**0.84**	**0.84**	**0.86**	**0.86**
	P-value	0.00	0.00	0.00	0.00	0.00
Growth stocks	Outperformance	−0.04	−0.02	−0.02	−0.01	−0.05
	P-value	0.68	0.57	0.59	0.54	0.68
Value stocks	Outperformance	−0.06	0.06	0.02	0.10	0.08
	P-value	0.74	0.27	0.43	0.17	0.21

Notes **BH** denotes the buy-and-hold strategy, whereas **COMBI** denotes the "combined" moving average trading strategy where at each month-end the best trading strategy in a back test is selected. The notations for the other trading strategies are self-explanatory. The out-of-sample period from January 1944 to December 2015. Outperformance is measured by $\Delta = SR_{MA} - SR_{BH}$ where SR_{MA} and SR_{BH} are the Sharpe ratios of the moving average strategy and the buy-and-hold strategy respectively. Bold text indicates the outperformance which is statistically significant at the 10% level

average trading strategy and the performance of the corresponding buy-and-hold strategy over bull and bear markets. The bull and bear markets are determined using the prices of the S&P Composite index. For each stock market index, Table 10.3 reports the descriptive statistics of the buy-and-hold strategy and the moving average trading strategy over bull and bear markets. The moving average strategy is simulated assuming monthly trading. The descriptive statistics include the mean and standard deviation of returns (in annualized terms), as well as the Sharpe ratios over the bull markets. The Sharpe ratios over the bear markets are not reported, because when the mean excess return is negative, the value of the Sharpe ratio is not reliable and hard to interpret.

Table 10.3 Descriptive statistics of the buy-and-hold strategy and the moving average trading strategy over bull and bear markets

Stock index	Statistics	Bull markets		Bear markets	
		BH	MA	BH	MA
DJIA	Mean returns %	22.79	15.77	−18.56	−8.98
	Std. deviation %	12.54	10.53	14.34	10.69
	Sharpe ratio	1.53	1.15		
Large stocks	Mean returns %	23.84	16.71	−21.47	−6.94
	Std. deviation %	12.33	10.74	14.15	9.46
	Sharpe ratio	1.64	1.21		
Small stocks	Mean returns %	29.53	21.98	−25.30	−7.84
	Std. deviation %	18.91	15.71	20.63	11.89
	Sharpe ratio	1.37	1.16		
Growth stocks	Mean returns %	24.51	17.30	−23.96	−10.81
	Std. deviation %	13.87	12.15	16.29	10.78
	Sharpe ratio	1.50	1.12		
Value stocks	Mean returns %	29.33	19.85	−19.62	−5.16
	Std. deviation %	16.11	13.32	17.72	10.08
	Sharpe ratio	1.59	1.21		

Notes **BH** and **MA** denote the buy-and-hold strategy and the moving average trading strategy respectively. Mean returns and standard deviations are annualized. Descriptive statistics are reported for the out-of-sample period from January 1944 to December 2015

Observe that, for each stock market index, over bull markets the buy-and-hold strategy outperforms the moving average trading strategy. Specifically, over bull markets the buy-and-hold strategy has both higher mean return and standard deviation compared to the moving average strategy. At the same time, over bull markets the buy-and-hold strategy has better risk-adjusted performance than the moving average strategy. In contrast, over bear markets the moving average trading strategy has better tradeoff between the risk and return than that of the buy-and-hold strategy. In particular, over bear markets the moving average strategy has substantially higher mean return (yet it is negative) with lower standard deviation compared to the buy-and-hold strategy. That is, even though in some stock markets the moving average strategy underperforms its passive counterpart on a risk-adjusted basis, in each stock market the properties of the moving average strategy resemble the properties of the portfolio insurance strategy which partially protects investors from losses during bear markets.

Finally in this section we would like to provide some words of caution regarding the superior performance of the moving average strategy in trading the small stocks. The first problem with trading the small stocks is that the small stocks are much less liquid as compared to the large stocks. As a result, the transaction costs in trading the small stocks are larger than those in trading the large stocks. In our simulations we used the same amount of transaction costs in each stock market. Therefore in our forward tests the outperformance produced by the moving average strategy in trading the small stocks is upward-biased. The second and much more serious problem in trading the small stocks is that the moving average strategy seems to have taken advantage of the existed strong short-term momentum in this market over the period from the early 1960 to the early 2000s. This short-term momentum ceased to exist and, consequently, the moving average strategy started to underperform the buy-and-hold strategy from the mid-2000s. Figure 10.2 plots the rolling outperformance delivered by the moving average strategy in daily trading the small stocks. The plot of the rolling out-of-sample outperformance of the moving average strategy resemble the plot of the rolling in-sample outperformance of the MOM(2) strategy depicted in Fig. 10.1. Since the moving

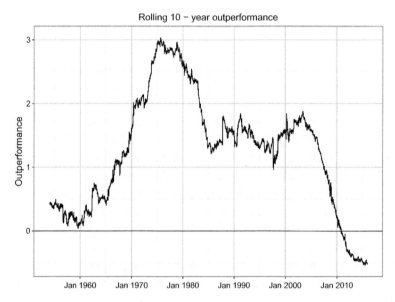

Fig. 10.2 Rolling 10-year outperformance in daily trading small stocks produced by the moving average strategy simulated out-of-sample over the period from January 1944 to December 2015. Outperformance is measured by $\Delta = SR_{MA} - SR_{BH}$ where SR_{MA} and SR_{BH} are the Sharpe ratios of the moving average strategy and the buy-and-hold strategy respectively

average strategy underperformed its passive counterpart over the course of the last decade, the chances that the moving average strategy will outperform the buy-and-hold strategy in the near future are very small in our opinion.

10.3   Bond Markets

10.3.1   Data

In this study, we use data on two bond market indices and the risk-free rate of return. These two bond market indices are the long-term and intermediate-term US government bond indices. Our sample period begins in January 1926 and ends in December 2011 (86 full years), giving a total of 1032 monthly observations. The bond data are from the Ibbotson SBBI 2012 Classic Yearbook. We use both the capital gain returns and total returns on long-term and intermediate-term government bonds. The trading signals are computed using bond index prices not adjusted for dividends. The risk-free rate of return is also from the Ibbotson SBBI 2012 Classic Yearbook. In particular, the risk-free rate of return for our sample period equals to 1-month Treasury Bill rate from Ibbotson and Associates Inc.

10.3.2   Bull and Bear Market Cycles in Bond Markets

Before testing the performance of the moving average strategies in the bond markets, it is useful to analyze the dynamics of the bull and bear market cycles in these markets. Figure 10.3, upper panel, plots the yield on the long-term US government bonds[4] over the period from January 1926 to December 2011, whereas the lower panel plots the natural log of the long-term government bond index over the same period. Shaded areas in the lower panel indicate the bear market phases. These bull and bear market phases are detected using the same algorithm as that used to detect the bull and bear market phases in the S&P Composite index.

The bull and bear markets depicted in Fig. 10.3, lower panel, are known as "primary markets"; the length of these markets generally lasts from one to five years in duration. Besides the primary market trends, one can easily observe the long-term trends known as "secular markets" or trends. A secular trend, that lasts from one to three decades, holds within its parameters many primary trends. For example, a secular bull market has bear market periods within it,

---

[4]The data are provided by Robert Shiller http://www.econ.yale.edu/~shiller/data.htm. Alternatively, these data can also be downloaded from https://www.measuringworth.com.

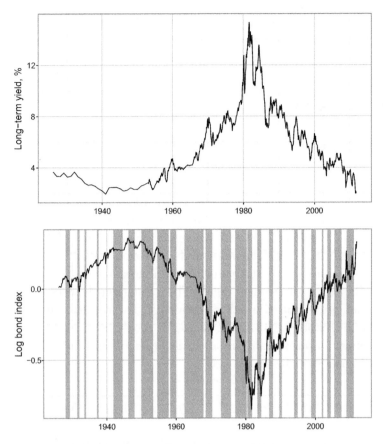

Fig. 10.3 The *upper panel* plots the yield on the long-term US government bonds over the period from January 1926 to December 2011, whereas the *lower panel* plots the natural log of the long-term government bond index over the same period. *Shaded areas* in the *lower panel* indicate the bear market phases

but it does not reverse the overlying trend of upward asset values. Similarly, a secular bear market has bull market periods within it.

Our historical sample contains two secular bull markets and one secular bear market. The secular bull markets cover the period from 1926 to the mid-1940s and the period from 1982 to 2011. These secular bull markets are associated with two long-term periods of decreasing yield on the long-term government bonds. The secular bear market spans the period from the mid-1940s to 1982 and is associated with a long-term period of increasing yield on the long-term government bonds. A visual investigation of the bull-bear cycles in the bond market suggests that the parameters and dynamics of these bull-bear cycles vary across secular markets. Specifically, a secular bull market is characterized by long bull and short bear primary trends. In contrast, a secular bear market is characterized by short bull and long bear primary trends. This knowledge suggests that we should expect that the performance of the moving average rules varies across secular bull and bear markets.

10.3.3 Back-Testing Trading Rules

For both intermediate- and long-term bonds, Table 10.4 shows the top 10 best trading strategies in a back test over the total period from January 1929 to December 2011, as well as over the two sub-periods: the first one is from January 1944 to December 1982 and the second one is from January 1983 to December 2011. We remind the reader that the first sub-period spans a secular bear market in bonds, whereas the second sub-period covers a secular bull market in bonds. The results reported in this table suggest the following observations. Over the whole in-sample period the best trading strategies

Table 10.4 Top 10 best trading strategies in a back test

Rank	Strategy	Δ	Strategy	Δ	Strategy	Δ
	1929–2011		1944–1982		1983–2011	
Long-term bonds						
1	SMAC(11,16)	0.08	SMAE(23,0.5)	0.40	SMAE(3,3.75)	−0.10
2	SMAC(11,15)	0.05	MOM(15)	0.40	MOM(13)	−0.11
3	MOM(13)	0.05	MOM(16)	0.39	SMAE(2,3.25)	−0.11
4	SMAE(4,5)	0.05	SMAE(22,0.75)	0.37	SMAE(3,5)	−0.12
5	SMAC(11,13)	0.03	MOM(18)	0.37	SMAE(4,5)	−0.12
6	SMAC(10,15)	0.03	SMAE(23,0.75)	0.37	SMAE(6,0.25)	−0.13
7	SMAC(10,16)	0.03	SMAE(22,0.5)	0.36	SMAC(11,15)	−0.13
8	SMAE(23,4.75)	0.03	SMAC(9,17)	0.36	SMAE(4,3.5)	−0.13
9	SMAC(9,17)	0.03	SMAE(24,0.5)	0.36	SMAC(2,6)	−0.13
10	MOM(15)	0.03	SMAC(4,20)	0.36	SMAE(4,2.5)	−0.13
Intermediate-term bonds						
1	SMAE(4,0.25)	0.13	MOM(15)	0.24	SMAE(2,2)	0.00
2	SMAC(8,16)	0.12	SMAE(4,0.25)	0.24	SMAE(2,2.25)	0.00
3	SMAC(8,15)	0.12	SMAC(7,20)	0.21	SMAE(2,2.5)	0.00
4	SMAE(2,1.75)	0.11	P-SMA(4)	0.21	SMAE(2,2.75)	0.00
5	P-SMA(5)	0.11	SMAC(11,17)	0.21	SMAE(2,3)	0.00
6	SMAC(8,19)	0.11	SMAC(10,18)	0.21	SMAE(2,3.25)	0.00
7	SMAC(8,17)	0.11	SMAE(2,1.5)	0.20	SMAE(2,3.5)	0.00
8	SMAC(8,18)	0.11	SMAE(2,1.75)	0.20	SMAE(2,3.75)	0.00
9	SMAC(9,15)	0.10	SMAC(5,16)	0.20	SMAE(2,4)	0.00
10	SMAC(7,20)	0.10	P-SMA(5)	0.20	SMAE(2,4.25)	0.00

Notes $\Delta = SR_{MA} - SR_{BH}$ where SR_{MA} and SR_{BH} are the Sharpe ratios of the moving average strategy and the buy-and-hold strategy respectively

outperform the buy-and-hold strategy on a risk-adjusted basis. However, the results for each sub-period advocate that the best trading strategies in a back test outperform the buy-and-hold strategy over the secular bear market only. In contrast, over the secular bull market in bonds, even the best trading strategies are not able to outperform the buy-and-hold strategy. In trading the long-term bonds, even the best trading strategy in a back test significantly underperforms the buy-and-hold strategy over the secular bull market in bonds.

10.3.4   Forward-Testing Trading Rules

The initial in-sample period in forward tests is from January 1929 to December 1943. Consequently, the out-of-sample period is from January 1944 to December 2011. Table 10.5 reports the (out-of-sample) outperformance delivered by the moving average strategies in trading the long- and intermediate-term bonds. The main conclusion that can be drawn from these results is clear-cut: the moving average rules do not outperform the buy-and-hold strategy in trading bonds. Specifically, for the majority of rules the outperformance is negative. For some rules the estimate for the outperformance is positive, but it is neither economically nor statistically significant. Besides, the outperformance is very uneven in time; the outperformance is positive mainly over the period that spans the secular bear market in bonds, see Fig. 10.4 for an illustration.

Whereas in stock markets the moving average rules provide significant downside protection (the maximum drawdown is reduced by approximately 50%), in bond markets the downside protection, as measured by the reduction in the maximum drawdown, amounts to only 25% which corresponds to the reduction in the mean excess returns. The fact that the dynamics of the bull-bear cycles in bond markets is changing over time suggests using walk-forward tests instead of forward tests. To test whether the outperformance is better in walk-forward tests, we simulated the returns to the moving average rules over the same out-of-sample period using a rolling in-sample window of 10 years. The results of these walk-forward tests are virtually the same as those of the forward tests (in order to save space, these results are not reported). Consequently,

Table 10.5 Descriptive statistics of the buy-and-hold strategy and the out-of-sample performance of the moving average trading strategies

| Statistics | BH | Moving average strategy | | | | |
		MOM	P-SMA	SMAC	SMAE	COMBI
Long-term bonds						
Mean returns %	6.28	5.83	5.37	5.81	5.37	5.64
Std. deviation %	9.08	6.50	6.55	6.56	6.37	6.39
Minimum return %	−11.24	−11.24	−11.24	−11.24	−11.24	−11.24
Maximum return %	15.23	14.43	11.45	14.43	11.45	14.43
Skewness	0.58	0.69	0.17	0.67	0.21	0.65
Kurtosis	4.12	10.75	7.60	10.03	7.95	10.92
Average drawdown %	3.98	3.07	2.71	3.24	2.94	3.30
Average max drawdown %	13.45	9.55	8.76	9.54	8.78	9.21
Maximum drawdown %	20.97	15.21	24.15	14.40	15.60	14.40
Outperformance		0.02	−0.05	0.02	−0.04	−0.00
P-value		0.41	0.69	0.42	0.68	0.52
Rolling 5-year Win %		31.97	54.95	35.80	34.08	31.44
Rolling 10-year Win %		44.48	64.13	37.30	49.35	29.27
Intermediate-term bonds						
Mean returns %	5.76	5.05	5.35	5.07	5.18	5.12
Std. deviation %	4.77	3.37	3.41	3.41	3.41	3.43
Minimum return %	−6.41	−6.51	−3.87	−3.87	−3.87	−3.87
Maximum return %	11.98	5.31	6.14	5.31	6.14	6.14
Skewness	0.88	−0.07	0.68	0.44	0.59	0.60
Kurtosis	7.76	6.96	5.91	4.75	5.95	5.75
Average drawdown %	1.64	1.46	1.23	1.43	1.42	1.38
Average max drawdown %	5.46	4.57	4.13	4.64	4.23	4.21
Maximum drawdown %	8.89	6.51	6.59	7.95	6.21	6.48
Outperformance		−0.07	0.02	−0.06	−0.03	−0.05
P-value		0.79	0.40	0.76	0.66	0.74
Rolling 5-year Win %		27.34	41.88	35.40	25.63	24.70
Rolling 10-year Win %		18.94	45.62	39.02	26.54	22.67

Notes **BH** denotes the buy-and-hold strategy, whereas **COMBI** denotes the "combined" moving average trading strategy where at each month-end the best trading strategy in a back test is selected. The notations for the other trading strategies are self-explanatory. Outperformance is measured by $\Delta = SR_{MA} - SR_{BH}$ where SR_{MA} and SR_{BH} are the Sharpe ratios of the moving average strategy and the buy-and-hold strategy respectively

even though the dynamics of the bull-bear cycles in bond markets is changing over time, walk-forward tests are not able to accommodate the parameters of moving average rules to the changing dynamics.

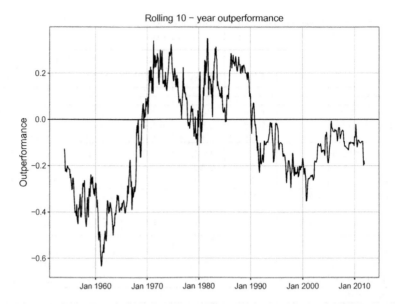

Fig. 10.4 Rolling 10-year outperformance in trading the long-term bonds produced by the moving average strategy simulated out-of-sample over the period from January 1944 to December 2011. Outperformance is measured by $\Delta = SR_{MA} - SR_{BH}$ where SR_{MA} and SR_{BH} are the Sharpe ratios of the moving average strategy and the buy-and-hold strategy respectively

10.4   Currency Markets

10.4.1   Exchange Rate Regimes

An exchange rate (a.k.a. a foreign-exchange rate, Forex rate, or FX rate) between two currencies is the rate at which one currency is exchanged for another currency. In simple terms, an exchange rate is the amount of a currency that one needs to pay in order to buy one unit of another currency.

An exchange-rate regime is the way a country's monetary authority, generally the central bank, manages its currency in relation to other currencies and the foreign exchange market. Before World War I, most countries adhered to the "gold standard". The countries that used the gold standard were committed to exchange their national currency in a fixed amount of gold. The gold standard creates a "fixed exchange" regime causing prices in different countries to move together, and hence create price stability. However, the gold standard had an adverse side-effect. Specifically, it put restrictions on a country's monetary policy. Without an increase in the amount of gold held as reserve in the country, government is not able to increase the money supply. An important implication is that a balance-of-payment deficit translates into a reduction in the gold reserve, which again translates into a reduction in the money supply (contractive monetary policy). In other words, if a country has a deficit in the balance-of-payments it must use a deflationary (change in the domestic price level) policy instead of a devaluation (change in the exchange rate). Countries with a balance-of-payment surplus could either do nothing or let the money supply increase, but then with the danger of creating inflation. This difference between countries with a deficit or surplus in the balance-of-payment creates an asymmetry in how the gold standard operates.

Most countries suspended the gold standard by the outbreak of World War I when the governments were in need of creating inflation in order to finance the war. After World War I, the gold standard was re-established. The classical gold standard ceased to exist because of the Great Depression and subsequent World War II. After World War II, a system similar to a gold standard and sometimes described as a "gold exchange standard" was established by the Bretton Woods Agreements. From 1946 to the early 1970s, the Bretton Woods system made fixed currencies the norm. The Bretton Woods system rested on both gold and the U.S. dollar. In principle, the system replaced the gold standard with the U.S. dollar. The countries, that were members of the Bretton Woods Agreements, agreed to redeem their currency for U.S. dollars, and the U.S. committed to exchange dollars for a fixed amount of gold.

Unfortunately, fixed exchange rates work satisfactory as long as the countries maintain their competitiveness and adhere to similar economic policies. Eventually the U.S. lost its competitiveness against Europe and Japan and the U.S. dollar became overvalued. The U.S. unilaterally terminated convertibility of the U.S. dollar to gold in 1971, effectively bringing the Bretton Woods system to an end. After that, "floating rates" are the most common exchange rate regime. Under the floating rates, a currency exchange rate depends on the supply and demand for this currency.

10.4.2   Data and Methodology

We consider a trader which home country is the U.S. Consequently, our convention is that we quote exchange rates as the price in U.S. dollars per unit of foreign currency (FC). That is, we quote the rate as USD/FC.

The exchange market consists of two core segments: the spot exchange market and the futures exchange market. The spot exchange market is the exchange market for payment and delivery of foreign currency "today". The futures exchange market is the exchange market for payment and delivery of foreign currency at some "future" date.

Our dataset consists of six spot exchange rates and seven 1- or 3-month government yields.[5] All data span the period from January 1971 to December 2015. Specifically, we obtain month-end exchange rates for Sweden, Japan, South Africa, Canada, the United Kingdom (U.K.), and Australia. To check whether there is any advantage in trading daily rather than monthly, we obtain day-end exchange rates for the U.K. and Australia. These data are obtained from the Federal Reserve Economic Data (FRED), a database maintained by the Research division of the Federal Reserve Bank of St. Louis.[6]

Under our convention, the trader buys foreign currency in the spot market.[7] The moving average trading rules are used to generate Buy and Sell signals. When the trading signal is Buy, the trader buys the foreign currency and deposits it in a foreign bank to earn the risk-free rate. When a Sell signal is generated, the trader converts the foreign currency to U.S. dollars and deposits them in a home bank to earn the risk-free rate. The risk-free rate of return in the U.S. equals to 1-month Treasury Bill rate provided by Ibbotson and Associates Inc. The risk-free rates of return in the six other countries equal to 1-month (or 3-months) Treasury Bill rates provided by the central banks in each country.

More formally, the currency capital gain return is computed as

$$r_t = \frac{X_t - X_{t-1}}{X_{t-1}},$$

where X_t denotes the end of period t exchange rate. The time series of $\{X_t\}$ is used to compute moving averages and generate trading signals. The total return on the foreign currency is the sum of the capital gain return and the foreign risk-free rate of return:

$$R_t = r_t + r^*_{f,t},$$

where $r^*_{f,t}$ denotes the foreign risk-free rate of return. The return on the U.S. dollar equals the domestic risk-free rate of return $r_{f,t}$.

Park and Irwin (2007) review, among other things, 38 studies where the researchers tested the profitability of technical trading strategies in currency markets. The great majority of these studies find profitability of technical trading strategies. However, several studies conducted in the early 2000s seem to suggest that technical trading profits have declined or disappeared since the mid-1990s (see Park and Irwin 2007, and references therein). The researchers

[5]These yields are proxies for the risk-free interest rates for the U.S. and six other countries.
[6]https://fred.stlouisfed.org/.
[7]A similar convention is used in, for example, Okunev and White (2003) and Kilgallen (2012).

jumped to the conclusion that the currency markets gradually became "efficient" which implies that it is not possible to "beat the market" consistently using the information from the past exchange rates. However, in our opinion this conclusion was premature. This is because in periods where the U.S. dollar strengthens (bulls market in the U.S. dollar), the market timing strategies do not work. The fact is that since the early 1970s the U.S. dollar tends to follow long-term (or secular) cycles lasting 5–10 years. The absence of profitability of technical trading rules over the period from the mid-1990s to the early 2000s can be explained by the fact that over this historical period the U.S. dollar was strengthening.

For the sake of illustration, Fig. 10.5 plots a weighted average of the foreign exchange value of the U.S. dollar against a subset of the broad index currencies.[8] These index currencies include the Euro Area, Canada, Japan, the United Kingdom, Switzerland, Australia, and Sweden. Shaded areas in this plot indicate the bear market phases. These bull and bear market phases are detected using the same algorithm as that used to detect the bull and bear market phases in the S&P Composite index. The graph in this plot advocates that there were two secular bull markets in the U.S. dollar. The first one lasted between 1980 and 1985, whereas the second one lasted between 1995 and 2003. The first secular bull market was induced by increasing interest rates in the U.S. (see the previous section on the bull and bear markets in bonds) and, as a consequence, high demand for the U.S. dollar. The second secular bull market covers the period of the Dot-Com bubble when the Internet and similar technology companies experienced meteoric rises in their stock prices and attracted substantial international capital flows.

It should be noted, however, that Fig. 10.5 plots a weighted average value of the U.S. dollar. Individual exchange rates may have own particular bull and bear cycles. Therefore when one tests the profitability of technical trading rules in some specific currency market, it makes sense to analyze the historical bull and bear market phases in this currency before jumping to a conclusion on whether technical trading rules work or do not work in this market.

10.4.3   Back-Testing Trading Rules

For each exchange rate, Table 10.6 shows the top 10 best (monthly) trading strategies in a back test over the period from January 1981 to December 2015. The results reported in this table suggest the following observations. First, for each exchange rate the best trading strategies outperform the

---

[8]The data for this plot are also obtained from the FRED database.

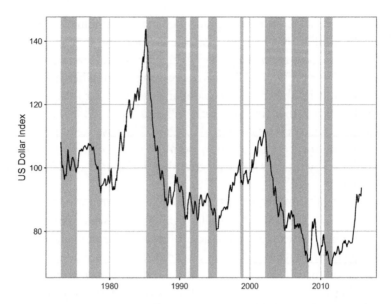

Fig. 10.5 A weighted average of the foreign exchange value of the U.S. dollar against a subset of the broad index currencies. *Shaded areas* indicate the bear market phases

buy-and-hold strategy on a risk-adjusted basis. The outperformance is the greatest in trading the US/Japan exchange rate and the lowest in trading the US/South Africa exchange rate. Second, for all exchange rates but the US/South Africa rate, the best trading strategies are based on exploiting a very short-term momentum in monthly rate. Specifically, for virtually all exchange rates the best trading strategies are either MOM(2) or SMAE(2,p) strategy. However, some trading strategies use a relatively long size of the averaging window. For example, in trading the US/Sweden exchange rate the familiar P-SMA(10) strategy shows the third best performance. Our third and final observation is that the SMAE(n, p) strategy is over-represented in the list of the top 10 strategies for all exchange rates.

For two exchange rates, Table 10.7 shows the top 10 best (daily) trading strategies in a back test over the same period from January 1981 to December 2015. Rather surprisingly, with daily trading the performance of the best trading strategies in a back test is significantly lower than that with monthly trading. This is surprising because daily trading in stock market indices produces better performance in back-tests as compared with monthly trading. The reasons for disadvantage of daily trading in currency markets lie in the time-series properties of exchange rates. In particular, for exchange rates the signal-to-noise ratio (see the preceding chapter for the definition of the signal-to-noise ratio) is lower than that for stock market indices. This is because the

Table 10.6 Top 10 best trading strategies in a back test with monthly trading

Rank	Strategy	Δ	Strategy	Δ	Strategy	Δ
	US/Sweden		US/Japan		US/South Africa	
1	MOM(2)	0.41	SMAE(2,3)	0.83	SMAE(6,5.5)	0.16
2	P-SMA(2)	0.41	SMAE(3,0.75)	0.77	SMAE(6,5.75)	0.16
3	SMAE(2,0.25)	0.37	SMAE(2,2.75)	0.76	SMAE(6,6)	0.16
4	P-SMA(10)	0.34	P-SMA(3)	0.75	SMAE(6,6.25)	0.16
5	SMAE(2,0.75)	0.34	SMAE(3,0.5)	0.74	SMAE(3,4.5)	0.16
6	SMAE(8,0.25)	0.33	SMAE(2,0.5)	0.73	SMAE(6,6.5)	0.16
7	SMAE(2,1.25)	0.33	SMAE(3,4.75)	0.72	SMAE(4,5.5)	0.15
8	SMAE(10,2)	0.32	SMAE(3,5)	0.72	SMAE(4,5.75)	0.15
9	SMAE(3,1.5)	0.32	SMAE(4,6)	0.72	SMAE(4,6)	0.15
10	SMAE(11,1.5)	0.32	SMAE(4,6.25)	0.72	SMAE(4,6.25)	0.15
	US/Canada		US/UK		US/Australia	
1	SMAE(2,1.5)	0.36	MOM(2)	0.45	MOM(2)	0.56
2	SMAE(4,2.75)	0.35	P-SMA(2)	0.45	P-SMA(2)	0.56
3	SMAE(4,3)	0.35	SMAE(2,1)	0.44	SMAE(2,0.25)	0.53
4	SMAE(2,1.25)	0.35	SMAE(2,0.25)	0.42	SMAE(2,0.5)	0.47
5	SMAE(7,4)	0.34	SMAE(2,2.25)	0.41	SMAE(3,0.25)	0.41
6	SMAE(8,4)	0.34	SMAE(5,0.5)	0.40	SMAE(6,3.75)	0.40
7	P-SMA(7)	0.34	P-SMA(3)	0.40	P-SMA(9)	0.40
8	SMAE(7,0.25)	0.33	SMAE(2,2)	0.40	P-SMA(8)	0.39
9	SMAE(6,0.75)	0.32	SMAE(8,2)	0.39	SMAE(3,0.5)	0.39
10	SMAE(5,1.75)	0.32	SMAE(4,2.25)	0.39	SMAE(8,0.25)	0.39

Notes $\Delta = SR_{MA} - SR_{BH}$ where SR_{MA} and SR_{BH} are the Sharpe ratios of the moving average strategy and the buy-and-hold strategy respectively

majority of exchange rates go sideways over a long run whereas stock market indices go up. In addition, while daily exchange rates exhibit very little or no persistence, monthly exchange rates show a high degree of persistence (see the concluding remarks to this chapter).

10.4.4 Forward-Testing Trading Rules

In these forward tests the initial in-sample period is from January 1974 to December 1983. Consequently, the out-of-sample period is from January 1984 to December 2015. Table 10.8 reports the outperformance delivered by the moving average strategies in out-of-sample tests in currency trading. For all exchange rates we simulate the out-of-sample returns to the moving average strategies assuming monthly trading. For 2 out of 6 exchange rates we also simulate the out-of-sample returns assuming daily trading. The results of these forward tests suggest the following observations. First, for 5 out of 6 exchange rates, the moving average strategies statistically significantly out-

Table 10.7 Top 10 best trading strategies in a back test with daily trading

Rank	Strategy	Δ	Strategy	Δ
	US/UK		US/Australia	
1	SMAE(12,4.5)	0.31	SMAE(210,8.5)	0.26
2	SMAE(13,4.75)	0.31	SMAE(220,8)	0.26
3	SMAE(170,3.25)	0.26	SMAE(200,8.75)	0.26
4	SMAE(140,3.5)	0.26	SMAE(220,8.25)	0.26
5	SMAE(160,2.25)	0.26	SMAE(220,8.5)	0.25
6	SMAE(130,3.75)	0.26	SMAE(80,7)	0.25
7	SMAE(190,0.5)	0.26	SMAE(220,8.75)	0.25
8	SMAE(130,4.25)	0.26	SMAE(230,8.25)	0.25
9	SMAE(150,3.25)	0.26	SMAE(190,9.25)	0.25
10	SMAE(140,4.25)	0.26	SMAC(13,170)	0.25

Notes $\Delta = SR_{MA} - SR_{BH}$ where SR_{MA} and SR_{BH} are the Sharpe ratios of the moving average strategy and the buy-and-hold strategy respectively

perform the buy-and-hold strategy in monthly trading. The best outperformance is achieved in trading the US/Japan exchange rate. Only in trading the US/South Africa exchange rate the outperformance is close to zero. Second, all moving average trading rules deliver about the same outperformance. That is, regardless of the choice of a trading rule, the out-of-sample performance of a moving average strategy remains virtually the same. Third, there is no advantage in trading daily rather than monthly. Specifically, in daily trading the outperformance is usually negative.

The analysis of the bull-bear markets in the US/Japan and US/South Africa exchange rates allows us to understand the reason for very good profitability of moving average rules in trading the US/Japan exchange rate and poor profitability of these rules in trading the US/South Africa exchange rate. Figure 10.6, left panel, plots the bull and bear market cycles in the US/Japan exchange rate, whereas the right panel in this figure plots the bull and bear market cycles in the US/South Africa exchange rate. Apparently, the moving average rules did not work in trading the US/South Africa exchange rate because the South African currency (South African Rand, ZAR) has been strengthening virtually over the whole out-of-sample period except some short historical episodes. In contrast, the moving average rules worked very well in trading the US/Japan exchange rate because the Japanese currency (Japanese Yen, JPY) has been weakening virtually over the whole out-of-sample period except some short historical episodes.

To get deeper insights into the properties of the out-of-sample performance of moving average trading rules in currency markets, Table 10.9 reports the detailed descriptive statistics of the buy-and-hold strategy and the out-of-sample performance of the moving average trading strategies in trading the

Table 10.8 Outperformance delivered by the moving average trading strategies in out-of-sample tests

FX Rate	Statistics	Moving average strategy				
		MOM	P-SMA	SMAC	SMAE	COMBI
Trading at the monthly frequency						
US/Sweden	Outperformance	**0.40**	**0.40**	**0.40**	**0.34**	**0.35**
	P-value	0.00	0.00	0.00	0.00	0.00
US/Japan	Outperformance	**0.62**	**0.73**	**0.73**	**0.66**	**0.65**
	P-value	0.00	0.00	0.00	0.00	0.00
US/South Africa	Outperformance	−0.08	−0.03	−0.05	0.03	0.03
	P-value	0.78	0.64	0.71	0.40	0.40
US/Canada	Outperformance	0.15	**0.28**	**0.23**	0.19	0.20
	P-value	0.18	0.02	0.04	0.14	0.12
US/UK	Outperformance	**0.32**	**0.32**	**0.32**	0.23	**0.31**
	P-value	0.01	0.01	0.01	0.06	0.02
US/Australia	Outperformance	**0.45**	**0.48**	**0.47**	**0.47**	**0.50**
	P-value	0.00	0.00	0.00	0.00	0.00
Trading at the daily frequency						
US/UK	Outperformance	−0.04	−0.04	−0.02	−0.04	−0.02
	P-value	0.89	0.89	0.78	0.86	0.70
US/Australia	Outperformance	0.02	−0.03	0.03	−0.02	−0.01
	P-value	0.31	0.78	0.22	0.69	0.65

Notes **BH** denotes the buy-and-hold strategy, whereas **COMBI** denotes the "combined" moving average trading strategy where at each month-end the best trading strategy in a back test is selected. The notations for the other trading strategies are self-explanatory. The out-of-sample period from January 1984 to December 2015. Outperformance is measured by $\Delta = SR_{MA} - SR_{BH}$ where SR_{MA} and SR_{BH} are the Sharpe ratios of the moving average strategy and the buy-and-hold strategy respectively. Bold text indicates the outperformance which is statistically significant at the 10% level

US/Sweden exchange rate. Observe that the standard deviation of the moving average trading strategies is only a bit less than that of the buy-and-hold strategy. However, the mean return to the moving average trading strategies is substantially higher than that of the buy-and-hold strategy. The moving average strategies are substantially less risky when risk is measured by the maximum drawdown(s). Therefore, in currency markets the moving average strategies are "high returns, low risk" strategies. Even though the outperformance is statistically significant, note that the outperformance is very uneven over time and there is absolutely no guarantee that over a 5- to 10-year period the moving average strategy outperforms its passive counterpart. To illustrate this feature of outperformance, Fig. 10.7, bottom panel, plots the 5-year rolling outperformance delivered by the combined moving average strategy, whereas the top panel in this figure plots the bull-bear markets in the US/Sweden exchange rate.

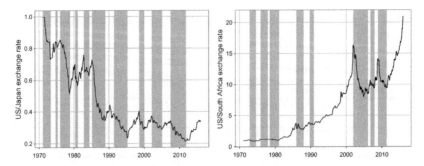

Fig. 10.6 *Left panel* plots the bull and bear market cycles in the US/Japan exchange rate. *Right panel* plots the bull and bear market cycles in the US/South Africa exchange rate. *Shaded areas* indicate the bear market phases

Table 10.9 Descriptive statistics of the buy-and-hold strategy and the out-of-sample performance of the moving average trading strategies in trading the US/Sweden exchange rate

		Moving average strategy				
Statistics	BH	MOM	P-SMA	SMAC	SMAE	COMBI
Mean returns %	6.23	10.74	10.87	10.87	9.74	9.87
Std. deviation %	9.14	8.81	8.81	8.81	8.89	8.88
Minimum return %	−6.84	−6.52	−6.52	−6.52	−7.28	−7.28
Maximum return %	12.73	12.73	12.53	12.53	12.53	12.53
Skewness	0.64	0.58	0.57	0.57	0.52	0.49
Kurtosis	2.01	1.97	1.90	1.90	1.98	1.98
Average drawdown %	6.79	3.07	3.00	3.00	3.56	3.33
Average max drawdown %	11.50	7.81	7.89	7.89	8.70	8.77
Maximum drawdown %	32.11	13.78	13.78	13.78	13.78	13.78
Outperformance		**0.52**	**0.53**	**0.53**	**0.40**	**0.42**
P-value		0.01	0.01	0.01	0.05	0.05
Rolling 5-year Win %		82.46	78.15	78.15	63.38	62.77
Rolling 10-year Win %		97.36	84.91	84.91	70.94	70.94

Notes **BH** denotes the buy-and-hold strategy, whereas **COMBI** denotes the "combined" moving average trading strategy where at each month-end the best trading strategy in a back test is selected. The notations for the other trading strategies are self-explanatory. Outperformance is measured by $\Delta = SR_{MA} - SR_{BH}$ where SR_{MA} and SR_{BH} are the Sharpe ratios of the moving average strategy and the buy-and-hold strategy respectively

The graphs in this figure suggest that the moving average strategy outperforms its passive counterpart only during bear markets.

As a final remark, it is worth noting the following. Neither in the stock nor in the bond market the moving average strategy with short selling the financial asset outperforms its counterpart where the trader switches to cash. However, our (unreported) results suggest that in currency markets for all exchange rates, but the US/South Africa exchange rate, the short selling strategy signifi-

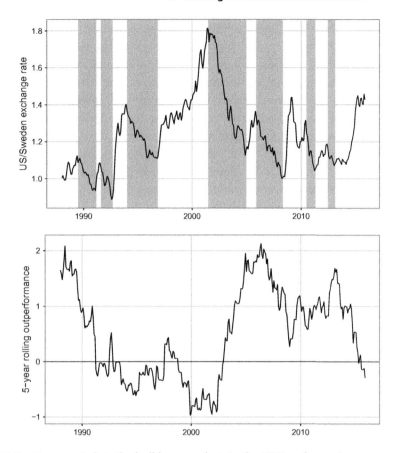

Fig. 10.7 *Top panel* plots the bull-bear markets in the US/Sweden exchange rate over the period from 1984 to 2015. *Shaded areas* indicate the bear market phases. *Bottom panel* plots the 5-year rolling outperformance delivered by the combined moving average strategy

cantly outperforms its counterpart in back-tests. In forward tests, on the other hand, for all exchange rates, but the US/Japan exchange rate, the short selling strategy only marginally outperforms its counterpart. Only for the US/Japan exchange rate the short selling strategy significantly outperforms its counterpart in forward-tests. However, this significant increase in outperformance is not surprising given the fact that the Japanese currency has been weakening virtually over the whole out-of-sample period and because the interest rate in the U.S. has been higher than that in Japan. In the subsequent paragraph we will elaborate more on the importance of these properties for the profitability of currency trading strategies.

In currency markets "short selling" a foreign currency means borrowing money in foreign currency, exchanging them to domestic currency, with subsequent saving in the domestic bank. This strategy is usually called a "currency carry trade" which consists in borrowing a currency at a low interest rate to finance the purchase of another currency earning a high interest rate. The idea of carry trade is to try to generate profits by exchanging two currencies with differing interest rates. In this regard, a currency carry trade can alternatively be called an "interest arbitrage". The practice of carry trade in currency markets gained popularity in the 1990s when there were large interest rate differentials between the economies in countries like Japan and the U.S. Specifically, at that time, interest rates in Japan had dropped to nearly zero, while rates in the U.S. were near 5% or above. A currency carry trade is risky because of the uncertainty in the exchange rate. The trader can lose money if the foreign currency appreciates. A moving average trading strategy can be used as an effective tool to hedge the currency carry trade risk and protect the trader from losses.

10.5   Commodity Markets

10.5.1   Historical Background

In economics, a commodity is "a marketable item produced to satisfy wants or needs". By a commodity one usually means a raw material or primary agricultural product that can be bought and sold. The price of a commodity is subject to supply and demand and inflation.

Commodity markets existed even in early civilizations. The Chicago Board of Trade (CBOT), established in the U.S. in 1848, is the first centralized financial market for trading futures contracts on commodities. The first traded contracts included such agricultural commodities as wheat, corn, cattle, and pigs. Since that time, the list of agricultural commodities has been considerable extended. Nowadays, besides various agricultural commodities, one can trade in futures contracts on energy (examples are crude oil and natural gas), metals (examples are copper and gold), raw materials (examples are timber and rubber), and fertilizers.

Financial econometric literature documents that, whereas stock prices are negatively correlated with inflation and interest rates (Fama 1981), commodity prices, on the other hand, are positively correlated with inflation and interest rates (Gorton and Rouwenhorst 2006; Kat and Oomen 2007). Commodity prices are also negatively correlated with stock prices. Therefore, when stock prices go down, commodity prices usually go up (Rogers 2007). To illustrate

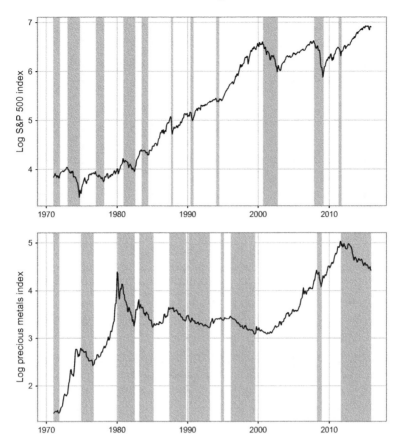

Fig. 10.8 *Top panel* plots the bull-bear cycles in the S&P 500 index over the period from 1971 to 2015. *Bottom panel* plots the bull-bear cycles in the Precious metals index over the same period. *Shaded areas* indicate the bear market phases

this feature of stock and commodity prices, Fig. 10.8, top panel, plots the bull-bear cycles in the S&P 500 index over the period from 1971 to 2015. The bottom panel in this figure plots the bull-bear cycles in the Precious metals index (gold, silver, and platinum) over the same period. The graphs in this figure suggest that when the stock prices go sideways (as in the 1970s and 2000s), the prices of precious metals increase substantially. Commodities had not been a popular asset class during the 1980s and 1990s. However, since the early 2000s when both the stock prices and interest rates started to decline, many investors have been attracted to commodities.

Investments in commodities often require a higher level of expertise to trade specific commodity futures contracts. To facilitate investment in commodity futures contracts, individual investors usually hire a Commodity Trading Advisor (CTA). There are three major investment styles employed by CTAs: technical, fundamental, and quantitative. According to various estimates, at least 2/3 of CTAs use technical analysis (trend following and momentum indicators) to make investment decision.

10.5.2 Data and Methodology

The World Bank has compiled monthly data that go back to 1960 for spot benchmark prices of a broad array of individual commodities and commodity price indices.[9] In our study we use commodity prices beginning from January 1971. This is because during the period of the gold exchange standard some commodity prices (for example, crude oil prices) exhibited little or no fluctuation. Only after the collapse of the Bretton Woods system and the abolishment of the gold exchange standard, all commodity prices began to fluctuate significantly. The abandonment of the gold standard made it possible for governments to use the banking system as a means to an unlimited expansion of money and credit. During the gold standard era, periods of inflation alternated with periods of deflation. On average, the prices remained on about the same level. The abandonment of the gold standard created a constant inflation without deflationary breaks. Even though there are significant differences between different commodities, since the early 1970s the average commodity prices have been steadily increasing.

Because of the big diversity of individual commodities, instead of testing the performance of moving average trading rules in each individual commodity market, we restrict our attention to testing these rules in 9 broad commodity price indices. The list of these indices and their components is presented in Table 10.10.

Even though we use commodity spot prices, we assume that when the trader buys and holds a commodity index, there are no costs of carry. That is, there are no costs of storing a physical commodity. This is equivalent of assuming that the trader buys short-maturity commodity futures contracts.[10] All commodity index prices are given in U.S. dollars. This also means that we consider the trader whose home country is the U.S. When the trader switches to cash, the return on cash equals to 1-month Treasury Bill rate provided by Ibbotson and Associates Inc.

---

[9]See http://www.worldbank.org/en/research/commodity-markets.

[10]The futures price converges to the spot price as the delivery date of the contract approaches, see any textbook on derivative securities, for example, Hull (2014).

Table 10.10 List of commodity price indices and their components

#	Commodity index	Components
1	Energy	Coal, Crude oil, and Natural gas
2	Beverages	Cocoa, Coffee, and Tea
3	Oils & Meals	Coconut oil, Fishmeal, Groundnuts, Palm oil, Soybeans, etc
4	Grains	Barley, Maize, Rice, Sorghum, and Wheat
5	Timber	Logs, Plywood, Sawnwood, and Woodpulp
6	Raw Materials	Cotton and Rubber
7	Fertilizers	DAP, Phosphate rock, Potassium chloride, TSP, and Urea
8	Base Metals	Aluminum, Copper, Lead, Nickel, Tin, and Zinc
9	Precious Metals	Gold, Platinum, and Silver

10.5.3   Back-Testing Trading Rules

For each commodity price index, Table 10.11 shows the top 10 best trading strategies in a back test over the period from January 1981 to December 2015. The returns to these strategies are simulated assuming that the trader switches to cash when a moving average rule generates a Sell signal. The following observations can be made. First, for all commodity price indices, the best trading strategies significantly outperform the buy-and-hold strategy on a risk-adjusted basis. Second, for all commodity indices but the Precious metals commodity index, the best trading strategies are based on exploiting a very short-term momentum in commodity prices. Specifically, for virtually all commodity price indices the best trading strategy is the $SMAE(2, p)$ strategy. Only in trading the Precious metals commodity index the best trading strategies use a relatively long size of the averaging window (from 7 to 14 months long). Our third and final observation is that the $SMAE(n, p)$ strategy is over-represented in the list of the top 10 strategies for all commodity indices but the Precious metals commodity index.

For all commodity price indices, the moving average strategy with short selling the commodity (when a Sell signal is generated) significantly outperforms its counterpart where the trader switches to cash (these results are not reported to save the space). Specifically, when short sales are allowed, in trading all commodity indices the outperformance increases from 30% to 100% in back-tests.

Table 10.11 Top 10 best trading strategies in a back test

Rank	Strategy	Δ	Strategy	Δ	Strategy	Δ
	Energy		**Beverages**		**Oils & Meals**	
1	SMAE(2,1.5)	0.57	SMAE(2,1.5)	0.59	SMAE(2,2.5)	0.59
2	SMAE(2,1)	0.55	SMAE(3,0.75)	0.58	SMAE(2,2.25)	0.58
3	SMAE(3,0.25)	0.52	SMAE(2,1.25)	0.58	SMAE(2,2)	0.57
4	SMAE(2,1.25)	0.52	SMAE(2,0.75)	0.58	SMAE(2,1.5)	0.57
5	SMAE(2,0.75)	0.51	SMAE(4,0.75)	0.57	SMAE(2,0.5)	0.56
6	SMAE(2,2)	0.51	SMAE(2,1)	0.57	SMAE(3,3)	0.56
7	P-SMA(3)	0.51	SMAE(4,1.25)	0.56	SMAE(3,2.75)	0.55
8	SMAE(3,0.75)	0.50	SMAE(3,1)	0.56	SMAE(3,3.75)	0.54
9	SMAE(3,0.5)	0.49	SMAE(4,1)	0.55	SMAE(6,3)	0.54
10	SMAE(3,1)	0.48	P-SMA(7)	0.55	SMAE(2,2.75)	0.53
	Grains		**Timber**		**Raw Materials**	
1	SMAE(3,0.25)	0.80	SMAE(2,0.25)	0.63	MOM(2)	1.05
2	SMAE(2,0.25)	0.79	SMAE(2,0.5)	0.62	P-SMA(2)	1.05
3	SMAE(2,0.5)	0.79	SMAE(3,0.25)	0.61	SMAE(2,0.25)	1.04
4	SMAE(3,0.5)	0.78	MOM(2)	0.60	SMAE(2,0.5)	0.98
5	P-SMA(3)	0.76	SMAE(2,1.25)	0.60	SMAE(2,0.75)	0.97
6	SMAE(2,0.75)	0.75	SMAE(3,0.75)	0.59	P-SMA(3)	0.96
7	SMAE(3,0.75)	0.75	SMAE(3,1)	0.57	SMAE(3,0.25)	0.94
8	MOM(3)	0.72	P-SMA(3)	0.57	SMAE(2,1)	0.94
9	SMAE(4,0.25)	0.71	SMAE(3,0.5)	0.56	SMAE(3,0.5)	0.93
10	P-SMA(4)	0.71	P-SMA(2)	0.56	SMAE(4,0.25)	0.91
	Fertilizers		**Base Metals**		**Precious Metals**	
1	SMAE(2,0.25)	0.89	SMAE(2,1.5)	0.67	SMAC(4,12)	0.50
2	P-SMA(2)	0.88	SMAE(2,2.5)	0.66	SMAC(3,13)	0.48
3	P-SMA(3)	0.88	SMAE(3,2.75)	0.65	SMAC(3,14)	0.48
4	MOM(2)	0.87	SMAE(3,2.25)	0.63	SMAC(4,13)	0.48
5	SMAE(2,0.5)	0.85	SMAE(4,2)	0.63	SMAE(11,2.75)	0.48
6	SMAE(3,0.25)	0.77	SMAE(3,2.5)	0.63	SMAE(12,1.25)	0.48
7	P-SMA(4)	0.74	SMAE(2,1.75)	0.63	SMAE(7,3)	0.47
8	P-SMA(5)	0.73	SMAE(6,0.25)	0.61	SMAE(13,1.25)	0.47
9	SMAE(5,0.5)	0.72	SMAE(4,1.75)	0.61	SMAC(6,13)	0.46
10	SMAE(3,0.5)	0.72	P-SMA(6)	0.61	SMAE(8,2.75)	0.46

Notes Short sales are not allowed. $\Delta = SR_{MA} - SR_{BH}$ where SR_{MA} and SR_{BH} are the Sharpe ratios of the moving average strategy and the buy-and-hold strategy respectively

10.5.4 Forward-Testing Trading Rules

In these forward tests the initial in-sample period is from January 1974 to December 1983. Consequently, the out-of-sample period is from January 1984 to December 2015. Table 10.12 reports the outperformance delivered by the moving average strategies in out-of-sample tests in trading commodity price indices. The out-of-sample returns are simulated assuming that the trader switches to cash when a moving average rule generates a Sell signal. The results

Table 10.12 Outperformance delivered by the moving average trading strategies in out-of-sample tests

Commodity index	Statistics	Moving average strategy				
		MOM	P-SMA	SMAC	SMAE	COMBI
Energy	Outperformance	**0.30**	**0.42**	**0.42**	**0.53**	**0.39**
	P-value	0.03	0.00	0.00	0.00	0.01
Beverages	Outperformance	**0.51**	**0.48**	**0.48**	**0.54**	**0.54**
	P-value	0.00	0.00	0.00	0.00	0.00
Oils & Meals	Outperformance	**0.48**	**0.48**	**0.48**	**0.40**	**0.38**
	P-value	0.00	0.00	0.00	0.00	0.00
Grains	Outperformance	**0.61**	**0.59**	**0.59**	**0.71**	**0.69**
	P-value	0.00	0.00	0.00	0.00	0.00
Timber	Outperformance	**0.55**	**0.52**	**0.39**	**0.51**	**0.44**
	P-value	0.00	0.00	0.00	0.00	0.00
Raw Materials	Outperformance	**1.00**	**0.96**	**0.96**	**0.95**	**0.92**
	P-value	0.00	0.00	0.00	0.00	0.00
Fertilizers	Outperformance	**0.80**	**0.62**	**0.62**	**0.79**	**0.75**
	P-value	0.00	0.00	0.00	0.00	0.00
Base Metals	Outperformance	**0.33**	**0.39**	**0.39**	**0.51**	**0.51**
	P-value	0.01	0.00	0.00	0.00	0.00
Precious Metals	Outperformance	**0.31**	**0.31**	**0.29**	0.17	**0.23**
	P-value	0.00	0.00	0.01	0.11	0.04

Notes **BH** denotes the buy-and-hold strategy, whereas **COMBI** denotes the "combined" moving average trading strategy where at each month-end the best trading strategy in a back test is selected. The notations for the other trading strategies are self-explanatory. The out-of-sample period from January 1984 to December 2015. Short sales are not allowed. Outperformance is measured by $\Delta = SR_{MA} - SR_{BH}$ where SR_{MA} and SR_{BH} are the Sharpe ratios of the moving average strategy and the buy-and-hold strategy respectively. Bold text indicates the outperformance which is statistically significant at the 10% level

of these forward tests suggest the following observations. First, for all commodity price indices, the moving average strategies statistically significantly outperform the buy-and-hold strategy (at the 10% level). The only exception is the outperformance of the SMAE rule in trading the Precious metals index; this outperformance is statistically significantly positive at the 11% level. Second, all moving average trading rules deliver about the same outperformance. That is, there is little variation in the outperformance across different moving average rules. Third, for all commodity price indices, the out-of-sample performance of the moving average trading rules increases from 20% to 50% when we allow short selling a commodity index (these results are unreported to save the space).

To get deeper insights into the properties of the out-of-sample performance of moving average trading rules in commodity markets, Table 10.13 reports the detailed descriptive statistics of the buy-and-hold strategy and the out-of-sample performance of the moving average trading strategies (with and without short sales) in trading the Grains index. Observe that when short sales are prohibited, the moving average strategy has significantly higher mean return and lower risk as compared to those of its passive counterpart. In particular,

Table 10.13 Descriptive statistics of the buy-and-hold strategy and the out-of-sample performance of the moving average trading strategies in trading the Grains commodity index

| Statistics | BH | Moving average strategy | | | | |
		MOM	P-SMA	SMAC	SMAE	COMBI
Short sales are prohibited						
Mean returns %	1.81	8.40	8.19	8.19	9.36	9.25
Std. deviation %	14.59	9.73	9.77	9.77	9.75	9.77
Minimum return %	−17.81	−9.39	−9.39	−9.39	−9.39	−9.39
Maximum return %	20.16	15.89	15.89	15.89	15.89	15.89
Skewness	0.42	1.41	1.38	1.38	1.46	1.46
Kurtosis	2.36	5.44	5.35	5.35	5.39	5.34
Average drawdown %	25.59	3.99	3.84	3.84	3.91	3.88
Average max drawdown %	25.59	10.06	9.38	9.38	8.26	8.26
Maximum drawdown %	57.09	20.07	20.89	20.89	17.85	17.85
Outperformance		**0.61**	**0.59**	**0.59**	**0.71**	**0.69**
P-value		0.00	0.00	0.00	0.00	0.00
Rolling 5-year Win %		92.62	92.31	92.31	92.62	92.62
Rolling 10-year Win %		100.00	100.00	100.00	100.00	100.00
Short sales are allowed						
Mean returns %	1.81	14.98	15.02	15.02	16.67	16.32
Std. deviation %	14.59	14.19	14.19	14.19	14.08	14.10
Minimum return %	−17.81	−19.08	−19.08	−19.08	−19.08	−19.08
Maximum return %	20.16	17.92	17.92	17.92	17.92	17.92
Skewness	0.42	0.11	0.11	0.11	0.12	0.13
Kurtosis	2.36	2.26	2.26	2.26	2.32	2.29
Average drawdown %	25.59	6.33	6.78	6.78	6.43	6.42
Average max drawdown %	25.59	15.55	15.86	15.86	14.03	14.33
Maximum drawdown %	57.09	28.68	28.68	28.68	24.73	27.77
Outperformance		**0.92**	**0.93**	**0.93**	**1.05**	**1.02**
P-value		0.00	0.00	0.00	0.00	0.00
Rolling 5-year Win %		94.46	92.92	92.92	93.23	93.23
Rolling 10-year Win %		100.00	100.00	100.00	100.00	100.00

Notes **BH** denotes the buy-and-hold strategy, whereas **COMBI** denotes the "combined" moving average trading strategy where at each month-end the best trading strategy in a back test is selected. The notations for the other trading strategies are self-explanatory. Outperformance is measured by $\Delta = SR_{MA} - SR_{BH}$ where SR_{MA} and SR_{BH} are the Sharpe ratios of the moving average strategy and the buy-and-hold strategy respectively

regardless of how the risk is measured, the riskiness of the moving average strategy is substantially lower as compared with the riskiness of the buy-and-hold strategy. For example, the standard deviation of returns of the moving average strategy is by 30% lower than the standard deviation of returns of the buy-and-hold strategy, whereas the maximum drawdown is lower by 60%. Allowing short sales enhances the mean return of the moving average strategy by approximately 40%. At the same time the standard deviation of the moving average strategy increases to a value comparable to the standard deviation of the buy-and-hold strategy; still the drawdowns of the moving average strategy remain on a significantly lower level as compared to the drawdowns of the buy-and-hold strategy. As in currency markets, the moving average strategy is "high returns, low risk" strategy.

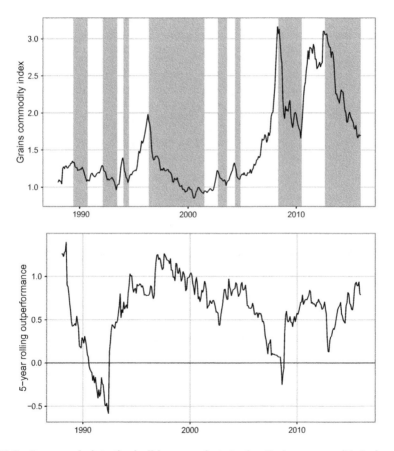

Fig. 10.9 *Top panel* plots the bull-bear markets in the Grains commodity index over the period from 1984 to 2015. *Bottom panel* plots the 5-year rolling outperformance delivered by the combined moving average strategy (where short sales are not allowed)

Figure 10.9, bottom panel, plots the 5-year rolling outperformance delivered by the combined moving average strategy, whereas the top panel in this figure plots the bull-bear markets in the Grains index. The graph of the rolling out-performance suggests that even though the outperformance varies over time, most of the time the outperformance remains positive. As a matter of fact, over a 10-year horizon the outperformance was always positive over our historical out-of-sample period (see Table 10.13, the estimates for the Rolling 10-year Win %).

10.6   Chapter Summary and Concluding Remarks

In this chapter we tested the performance of the moving average trading strate-gies in different financial markets: stocks, bonds, currencies, and commodities. The results of our tests allow us to draw the following conclusions:

- The moving average trading strategies performed best in commodity mar-kets. The next best performance of these strategies was observed in currency markets. The moving average strategies did not work in bond markets. It should be noted, however, that our sample of historical data for both the currencies and commodities was much shorter than that for both the stocks and bonds. Therefore one can question whether the historical sample for currencies and commodities is a truly representative sample from which one can draw reliable conclusions.
- In stock markets the outperformance produced by a moving average trad-ing strategy depends on the type of the stock price index. Statistically significant long-run outperformance was observed only in trading the small stock index; yet over the recent past the trading in small stocks became unprofitable. Trading in a well-diversified portfolio of large stocks seems to produce a robust long-run outperformance. However, this outperformance is not statistically significant at the conventional statistical levels.
- Outperformance delivered by a moving average trading strategy is very uneven regardless of the type of financial market. Consequently, over a short-run there is absolutely no guarantee for outperformance even in commodity markets.
- Short selling strategy is beneficial mainly in the commodity markets. In stock and bond markets the moving average strategy with shorts sales is very risky and significantly underperforms the buy-and-hold strategy.

- Regardless of the financial market, there is no advantage in trading daily rather than monthly. Only with daily trading the small cap stocks the out-of-sample performance of the moving average strategy was substantially better than that with monthly trading. However, even in this case the profitability of the moving average strategy disappeared in the recent past.

- Regardless of the financial market, the moving average strategy outperforms its passive counterpart basically over relatively long bear markets. Conversely, over bull markets the moving average strategy underperforms the buy-and-hold strategy. In each market there are secular trends that last from 10 to 30 years. Consequently, the moving average strategy might underperform its passive counterpart even over a long-run. For example, in stock markets the moving average strategy underperformed the buy-and-hold strategy over the secular bull market in stocks that lasted from 1982 to 2001.

- Among all tested rules (MOM, SMAC, and SMAE), the SMAE rule (which generalizes the P-SMA rule) usually performs the best in the majority of markets and data frequencies. Therefore if the trader wants to use only a single trading rule, the SMAE rule should be preferred.

In addition to the set of conclusions, the results of our tests suggest the following practical recommendations for traders testing the profitability of moving average trading rules:

- It is important to control the robustness of the historical outperformance delivered by a moving average trading strategy. Just looking at the estimate for the historical outperformance is not enough to jump to the conclusion that the strategy is profitable, because this estimate is related to the average outperformance over a rather long run. The outperformance is usually very uneven in time, but it should be positive over bear markets. Most importantly, the trader should control that the outperformance is positive over the most recent bear markets. This is needed because the market's dynamics can change and, as a possible consequence, the outperformance might disappear.

- The profitability of moving average trading rules in a financial market can be roughly evaluated by analyzing the historical dynamics of the bull and bear cycles in this market. The quantity of major interest is the ratio of the average bull market length to the average bear market length. If this ratio is close to or less than 1 (as in the majority of currency and commodity markets), the chances that the moving average rules outperform the buy-and-hold strategy are very high. If, on the other hand, this ratio is close to or greater than 2, the chances that the moving average rules outperform the buy-and-hold strategy are rather small. In this case the trader can only hope that the moving average strategy provides a superior downside protection during severe bear markets.

Our results also suggest the hypothesis that in a financial market there might exist both a short- and a long-term trend in prices. As a matter of fact, the hypothesis about simultaneous existence of several trends with different durations is not new in technical analysis. For example, Charles Dow, who is considered the father of modern technical analysis, developed a theory, later called the Dow Theory, which expresses his views on price actions in the stock market. Among other things, Dow Theory postulates that a market has three movements: the "main movement" (primary trend), the "medium swing" (secondary trend), and the "short swing" (minor trend). The three movements may be simultaneous, for instance, a minor movement in a bearish secondary reaction in a bullish primary movement.

The Dow Theory might be wrong, but the hypothesis about simultaneous existence of several trends, or momenta, in asset prices might be fruitful all the same. In finance, momentum denotes the empirically observed tendency for rising asset prices to keep rising, and falling prices to keep falling. The presence of different momenta in asset prices can be revealed, for example, by examining the first-order autocorrelation function (AC1) of k-day returns. To compute this first-order autocorrelation function, one can regress k-day returns on lagged k-day returns (this idea is presented in the famous paper by Fama and French 1988). Formally, one runs the following regression

$$\sum_{i=0}^{k-1} r_{t+i} = a(k) + b(k) \sum_{i=1}^{k} r_{t-i} + \varepsilon_t, \tag{10.1}$$

where r_t denotes the natural log of the day t asset return and k is varied between 1 and 250. The slopes of the regression, $b(k)$, are the estimated autocorrelations of k-day returns, $AC1(k)$. If the prices follow a Random Walk, there is no relationship between future and past returns and, consequently, $b(k) = 0$.[11] Evidence of momentum (mean-reversion) in asset prices comes from the positive (negative) values of $b(k)$. The larger the absolute value of $b(k)$, the stronger the momentum (or mean reversion if $b(k)$ is negative) in asset prices.

---

[11] However, if a sample is rather short, the estimate for $b(k)$ is downward biased. Therefore, even if asset prices follow a random walk, in short samples $b(k) < 0$ and decreases as k increases.

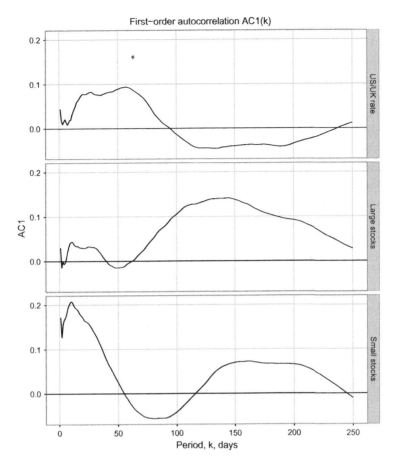

Fig. 10.10 Empirical first-order autocorrelation functions of k-day returns in the following financial markets: the US/UK exchange rate, the large cap stocks, and the small cap stocks

Figure 10.10 plots the empirical first-order autocorrelation functions of k-day returns in several financial markets. These markets are: the US/UK exchange rate, the large cap stocks, and the small cap stocks. The graphs in this figure suggest the following observations. First, in the US/UK exchange rate, there was only a short-term momentum over periods of 20–60 days. This says, for instance, that if the US/UK exchange rate had increased (decreased) over the course of the previous 30 trading days, this rate would tend to increase (decrease) further over the subsequent 30 trading days. Second, in both large- and small cap stocks there were two momenta: one short-term momentum over periods of 5–30 days, and one long-term momentum over periods of 150–200 days. In large stocks, the long-term trend was much stronger than the short-term trend. Conversely, in small stocks the short-term trend was much stronger than the long-term trend. We remind the reader that our tests revealed that, in the absence of transaction costs, in all stock markets the best trading strategy in a back test was the MOM(2) strategy that exploits the short-term momentum in stock prices. Even in the presence of realistic transaction costs, the best trading strategy in small stocks exploited also the short-term momentum. However, this short-term momentum ceased to exist in stock prices; anyway the long-term momentum seems to remain.

Simultaneous existence of several trends has significant practical implications. Specifically, all existed forward tests of profitability of technical trading rules are designed to find a single trend: the one that produces the best observed performance (of a moving average strategy) in a back test. In the presence of two trends of different durations, a profitable and robust trading strategy can in principle exploit both trends. An example of such strategy can be found in the paper by Glabadanidis (2017). The strategy in this paper is a sheer example of an ad-hoc strategy presented without any justification, but in some miraculous way the strategy is able to produce a superior performance. We argue that the superior performance of this strategy can be explained by the presence of both a short- and a long-term momentum in stock prices.

Simultaneous existence of several trends with different durations is able to explain a major controversy among technical traders about the optimal size of the averaging window in each trading rule. Even for the famous P-SMA rule, the popular advice on the length of the averaging window varies from 10 to 200 days (Kirkpatrick and Dahlquist 2010, Chap. 14). Apparently, different sizes of the averaging window appear because of the existence of trends of different durations; the duration of a trend varies over time and across different financial markets.

Finally in this chapter we would like to present the key descriptive statistics of returns on different financial asset classes: stocks, bonds, currencies, commodities, and cash. Table 10.14 reports the descriptive statistics of both the buy-and-hold strategy and the moving average trading strategy (the combined strategy simulated out-of-sample). As in the preceding chapter, we use 2-year returns instead of monthly returns. This is because monthly returns are not able to properly convey the idea of downside protection delivered by the moving average trading strategy. The descriptive statistics for all asset classes are computed for the 25-year period from January 1986 to December 2011. The reason for using such a short historical sample is because the data on currencies start only from January 1973 (the period from January 1976 to December 1985 is used as the initial in-sample period for simulating the returns to the moving average trading strategy). The different asset classes in this table exhibit

Table 10.14 Descriptive statistics of 2-year returns on different financial asset classes over 1986–2011

	Stocks		Bonds		Currencies		Commodities		Cash
	BH	MA	BH	MA	BH	MA	BH	MA	Cash
Mean return, %	22.49	20.38	14.66	12.41	11.01	13.23	15.55	22.76	8.40
Std. deviation, %	28.45	22.66	7.04	6.86	10.78	5.99	26.58	17.75	4.42
Skewness	−0.25	0.60	0.11	0.04	−0.11	0.39	0.42	0.37	−0.15
Probability of loss, %	19.38	16.96	0.35	3.81	16.96	0.00	33.22	9.34	0.00
Expected loss if loss occurs, %	−22.04	−9.55	−0.26	−1.89	−5.74		−11.87	−6.13	

Notes BH denotes the buy-and-hold strategy, whereas MA denotes the moving average strategy. The descriptive statistics are computed using the data over the period from January 1986 to December 2011. Stocks are the large cap stocks. Bonds are the intermediate-term government bonds. Currencies is a weighted average of the foreign exchange value of the U.S. dollar against a subset of the broad index currencies. Commodities is a weighted average of all commodities in the World Bank database except the precious metals. The return on Cash is proxied by the 1-month TBill rate

either small or no correlation with each other; therefore they can be used for efficient portfolio diversification across different asset classes.

Our first observation is that the moving average strategy does not work in the bond markets. Specifically, the moving average strategy is less rewarding and more risky than the corresponding buy-and-hold strategy in bonds. Even though the sample period 1986–2011 covers a single secular bull market in bonds, over a much longer historical sample from 1929 to 2011 the descriptive statistics of the moving average strategy in trading bonds are virtually the same. Our second observation is that in the stock market the moving average trading strategy is less rewarding, but at the same time substantially less risky when the risk is measured by the probability of loss and expected loss. Our third observation is that in both the commodity and currency markets the moving average trading strategy is both more rewarding and less risky than the corresponding buy-and-hold strategy. Interestingly, in currency trading the moving average strategy produced risk-free returns over the period 1986–2011 when the risk is measured by the probability of loss and expected loss. Besides, the risk-free returns to this strategy were substantially higher than the returns on risk-free cash. Last but not least, the mean return to the moving average strategy in currency trading is comparable to the mean return on the passive bond investing. The moving average strategy in commodity trading produced higher mean returns with lesser risk as compared to either passive or active investment in stocks. However, the reader should be reminded that the 25-year sample period is probably not a long enough and truly representative sample from which one can draw reliable conclusions about the performance of moving average rules in both commodity and currency markets.

References

Fama, E. F. (1981). Stock returns, real activity, inflation, and money. *American Economic Review, 71*(4), 545–565.

Fama, E. F., & French, K. R. (1988). Permanent and temporary components of stock prices. *Journal of Political Economy, 96*(2), 246–273.

Glabadanidis, P. (2017). Timing the market with a combination of moving averages (forthcoming in International Review of Finance).

Gorton, G., & Rouwenhorst, K. (2006). Facts and fantasies about commodity futures. *Financial Analysts Journal, 62*(2), 47–68.

Hull, J. C. (2014). *Options, futures, and other derivatives* (9th edn.). Pearson.

Kat, H. M., & Oomen, R. C. (2007). What every investor should know about commodities part II: Multivariate return analysis. *Journal of Investment Management, 5*, 40–64.

Kilgallen, T. (2012). Testing the simple moving average across commodities, global stock indices, and currencies. *Journal of Wealth Management, 15*(1), 82–100.

Kirkpatrick, C. D., & Dahlquist, J. (2010). *Technical analysis: The complete resource for financial market technicians* (2nd edn.). FT Press.

Okunev, J., & White, D. (2003). Do momentum-based strategies still work in foreign currency markets? *Journal of Financial and Quantitative Analysis, 38*(2), 425–447.

Park, C.-H., & Irwin, S. H. (2007). What do we know about the profitability of technical analysis? *Journal of Economic Surveys, 21*(4), 786–826.

Rogers, J. (2007). *Hot commodities: How anyone can invest profitably in the world's best market.* Random House Incorporated.

11

Conclusion

Besides providing the in-depth coverage of various types of moving averages, their properties, and technical trading rules based on moving averages, this book offers two new contributions to the field of technical analysis of financial markets. Specifically, this book uncovers the anatomy of market timing rules with moving averages of prices and performs the objective tests of profitability of moving average trading rules in different financial markets. In the concluding chapter we would like to summarize the two main contributions and make additional useful remarks regarding their significance.

11.1 Anatomy of Trading Rules

We considered the computation of the value of a technical indicator in all trading rules and showed that this value is computed using the past n closing prices including the last closing price

$$\text{Indicator}_t^{TR(n)} = f(P_t, P_{t-1}, \ldots, P_{t-n+1}),$$

where TR denotes a trading rule, P_{t-i} denotes the period $t - i$ closing price, and $f(\cdot)$ denotes the function that specifies how the value of the technical trading indicator is computed. In the original formulation, $f(\cdot)$ is a function of one or multiple moving averages of prices. This function is sometimes rather intricate which makes it difficult to comprehend how a given trading rule differs from the others. In the absence of understanding how a trading rule works, traders are more likely to have superstitious beliefs about the performance of

© The Author(s) 2017
V. Zakamulin, *Market Timing with Moving Averages*, New Developments
in Quantitative Trading and Investment, DOI 10.1007/978-3-319-60970-6_11

complex trading rules; believing in "the more complex, the better" is a common cognitive bias.

Our analysis demonstrates that the computation of a technical trading indicator for every moving average trading rule can alternatively be given by the following simple formula

$$\text{Indicator}_t^{TR(n)} = \sum_{i=1}^{n-1} w_i \Delta P_{t-i}. \tag{11.1}$$

where $\Delta P_{t-i} = P_{t-i+1} - P_{t-i}$ denotes the price change from $t - i$ to $t - i + 1$ and w_i is the weight of price change ΔP_{t-i} in the computation of a weighted moving average of price changes. Despite a great variety of trading indicators that are computed seemingly differently at the first sight, we found that the only real difference between the diverse trading indicators lies in the weighting function used to compute the moving average of price changes. The most popular trading indicators employ either equal-weighting of price changes, overweighting the most recent price changes, a hump-shaped weighting function which underweights both the most recent and most distant price changes, or a weighting function that has a damped waveform where the weights of price changes periodically alter sign.

We derived several closed-form solutions for the weights w_i of some trading rules coupled with the ordinary moving averages. It is a daunting task to derive closed-form solutions for the weights w_i for all existing trading rules and types of moving averages. Besides, in some cases it might not be possible to obtain a closed-form solution for the weights. Fortunately, there's a simple way around this problem. Specifically, since our main result tells us that the value of a trading indicator is a weighted average of past price changes, one can easily recover the weights by computing the value of the technical indicator using function $f(\cdot)$ and then regressing this value on the past price changes. That is, after computing the series of $\text{Indicator}_t^{TR(n)}$, one can run the following regression:

$$\text{Indicator}_t^{TR(n)} = \alpha + w_1 \Delta P_{t-1} + w_2 \Delta P_{t-2} + \ldots + w_{n-1} \Delta P_{t-n+1} + \varepsilon_t.$$

The estimated regression coefficients w_i represent the empirical weights of the price changes in the computation of the given trading indicator.

Let us elaborate further on the alternative representations of our main result on the anatomy of trading rules given by Eq. (11.1) and the intuition that can be gained from these representations. Since the positive (negative) value of the technical indicator predicts a price increase (decrease) over the subsequent period,[1] Eq. (11.1) can be rewritten as

$$\text{sgn}\,(\Delta P_t) = \sum_{i=1}^{n-1} w_i \Delta P_{t-i},\tag{11.2}$$

where sgn($\cdot$) is the mathematical sign function. Consequently, every trading rule can be interpreted as a predictive linear relationship between the weighted sum of past price changes and the direction of the future price change.

If the prices are defined in terms of logarithmic prices, then the difference between two successive log prices gives the logarithmic return. Formally, $r_t = P_t - P_{t-1}$, where P denotes the log price level and r_t denotes the log return. Therefore Eq. (11.2) can be rewritten as

$$\text{sgn}\,(r_{t+1}) = \sum_{i=0}^{n-2} w_i\, r_{t-i}.\tag{11.3}$$

In words, every trading rule can be interpreted as a predictive linear relationship between the weighted sum of past (log) returns and the sign of the future (log) return. Equation (11.3) is also approximately valid if one uses arithmetic returns instead of logarithmic returns. To see this, let us divide the left- and right-hand sides of Eq. (11.2) by P_t. This yields

$$\text{sgn}\left(\frac{\Delta P_t}{P_t}\right) = \sum_{i=1}^{n-1} w_i \frac{\Delta P_{t-i}}{P_t}.\tag{11.4}$$

The fraction $\frac{\Delta P_t}{P_t}$ gives the arithmetic return from t to $t+1$ which is denoted by r_{t+1}. Consider the fraction

$$\frac{\Delta P_{t-1}}{P_t} = \frac{\Delta P_{t-1}}{P_{t-1} + \Delta P_{t-1}}.$$

---

[1]For example, a positive value of a technical indicator generates a Buy signal for the next period. This Buy signal tells the trader that the prices trend upward and this trend will persist in the near future. Therefore, a Buy signal predicts that over the next period the price will increase.

If the prices are observed at a daily or monthly frequency, the average one-period change in the price is less than 1% which means that in the majority of cases $\Delta P_{t-1} \ll P_{t-1}$. Therefore the given fraction can be closely approximated by

$$\frac{\Delta P_{t-1}}{P_t} \approx \frac{\Delta P_{t-1}}{P_{t-1}} = r_t.$$

The same reasoning can be applied to all fractions $\frac{\Delta P_{t-i}}{P_t}$. Consequently, when the returns are defined in terms of the ordinary (arithmetic) returns, we can still rewrite Eq. (11.2) using returns instead of price changes. That is, Eq. (11.3) is approximately valid if one uses arithmetic returns.

It is worth emphasizing once more that Eq. (11.3) is none other than the predictive linear relationship between the weighted sum of the past returns and the sign of the future return. If we want to estimate empirically the weights in this predictive relationship, we can run the following regression (since n is an arbitrary integer value, without the loss of generality and for simplicity, we replace $n - 2$ with just n):

$$\text{sgn}\,(r_{t+1}) = \alpha + \sum_{i=0}^{n} w_i\, r_{t-i} + \varepsilon_t. \tag{11.5}$$

A closer look at regression (11.5) reveals that this regression resembles many models in modern empirical finance. The only difference is that in empirical finance it is more common to predict the future return instead of the sign of the future return. For example, the regression

$$r_{t+1} = \alpha + \sum_{i=0}^{n} \beta_i\, r_{t-i} + \varepsilon_t \tag{11.6}$$

is a familiar Auto-Regressive model of order n (AR(n) model) presented by Box and Jenkins (1976). The simplest form of this models is AR(1) model that has been extensively used in finance econometric literature over the course of the last 40 years

$$r_{t+1} = \alpha + \beta r_{t-1} + \varepsilon_t. \tag{11.7}$$

In this model, if coefficient β is statistically significantly different from zero, we have evidence of momentum (mean-reversion) when $\beta > 0$ ($\beta < 0$). In the presence of (one-period) momentum, the MOM(2) strategy is usually profitable, at least in the absence of transaction costs.

To predict future returns, Jegadeesh (1991) used the following predictive regression

$$r_{t+1} = \alpha + \beta \sum_{i=0}^{n} r_{t-i} + \varepsilon_t, \tag{11.8}$$

which represents a specific form of regression (11.6) where all past returns are equally weighted (in fact, this regression is virtually equivalent to the MOM(n) trading indicator). Fama and French (1988), and many other researchers afterwards, used another predictive regression

$$\sum_{i=1}^{n} r_{t+i} = \alpha + \beta \sum_{i=0}^{n} r_{t-i} + \varepsilon_t. \tag{11.9}$$

Whereas in regression (11.8) the sum of past n returns is used to predict the next period return, in regression (11.9) the sum of past n returns is used to predict the return over the subsequent n periods.

Overall, our result on the anatomy of moving average trading rules can be re-stated in terms of a predictive linear relationship between a weighted sum of past returns and the sign of the future return. This predictive linear relationship resembles very closely the models that have been used in empirical finance literature. Therefore our result allows us to reconcile modern empirical finance with technical analysis of financial markets that uses moving average rules, because both these approaches employ, in fact, the same type of a predictive linear model. Much of the academic criticism of technical analysis is focused on the Efficient Market Hypothesis, which states, even in its "weak form", that financial asset prices follow a Random Walk; therefore past prices cannot be used to predict future prices. At the same time, financial researchers have discovered evidence that prices do not follow a Random Walk. Specifically, prices exhibit momentum (see, for example, Jegadeesh and Titman 1993; Moskowitz et al. 2012) and mean-reversion (see, for example, Jegadeesh 1991, and Balvers et al. 2000). Consequently, if prices exhibit momentum, then the trader can try to profit from using this momentum. Regarding the technical trading with moving averages, it cannot be said that it is nonsense, because the core idea in this market timing technique is to profit from either momentum or mean-reversion which existence is documented in numerous financial studies. Still, the critique that the technical trading with moving averages represents a pseudo-science is warranted, but only because the majority of claims about profitability of the moving average trading rules are not supported by objective scientific evidence. However, all these claims are testable and the current challenge is to perform objective scientific tests of all these claims.

11.2   Profitability of Trading Rules

We assessed the profitability of moving average trading rules in different financial markets: stocks, bonds, currencies, and commodities. Our results showed a clear superiority of moving average trading rules in the currency and commodity markets.[2] In most of currencies and commodities in our study, the best performing strategy is based on using a relatively short-term momentum (or persistency) in prices over horizon of one month. This corresponds very well with the practical recommendations of using a 20- to 30-day moving average of daily prices to profit in commodity markets, see Kleinman (2005). The existence of a short-term momentum in these markets should not be surprising. Consider, for example, a commodity price which depends upon demand and supply of this commodity and inflation rate. Both the demand and supply of a commodity, as well as the inflation rate, are rather persistent over a short-run. Similarly, an exchange rate depends upon supply and demand of currency, interest rates in the respective countries, and some other macro-economic variables. All of these processes exhibit a short-term persistency. For example, the interest rate in a country is regulated by the central bank; yet the central bank revises the level of interest rates once in a few months only.

Our results suggest that moving average rules did not work in the bond markets; therefore it is unlikely that these rules will work in the bond markets in the future. The results for the stock markets are probably the most intriguing among all our results. First of all, we did not find a clear-cut answer to the question of whether the moving average trading strategy is superior to the buy-and-hold strategy. Yet, our results are encouraging even though they are in sharp contrast with those reported in the majority of previous studies where the authors claim that "one can easily beat the market using moving averages" and moving averages "allow one both to enhance returns and reduce risk at the same time". We found that the profitability of moving averages is highly overstated, to say the least. In other words, moving averages do not offer a quick and easy way to riches. On the other hand, moving average rules can protect from losses when this protection is most needed. Specifically, during a period of a severe market downturn when stock prices are trending downwards over a relatively long run, the moving average strategy mandates to switch to cash and, therefore, limits the losses. However, this downside protection comes at the expense of lowering the returns during the good states of the market.

---

[2]It should be emphasized, however, that in our study we used the spot commodity prices and exchange rates. In real trading, on the other hand, one uses futures contracts. We conjecture that in trading futures contracts the results are about the same, yet our conjecture is nothing more than that at this point, and will not be anything more until somebody validates it using historical data for futures prices.

In our opinion, the moving average trading strategy represents a prudent investment strategy for "moderate" and even "conservative" medium- and long-term investors. However, our results reveal that not every stock market index is suitable for timing the market using moving averages. The most robust performance of the moving average trading rules is observed when the stock market index represents a well-diversified portfolio of large cap stocks; a well-known example of such index is the S&P 500 index.

The practitioners find it comforting to know that the popular strategy that uses a 10-month SMA is close to the best performing rule for timing the S&P 500 index. As compared with the MOM(n) rule, the performance of the P-SMA(n) rule is much more robust with respect to the change in the size of the averaging window, n. For example, when the value of n varies in between 6 and 14, the performance of the MOM(n) rule changes much more significantly than the performance of the P-SMA(n) rule. This is because the duration of momentum in stock prices varies over time: in some periods the best accuracy of forecasting the future returns is attained when one uses the returns over the past 3–5 months, in other periods one needs to use returns over the past 10–16 months. The MOM(n) rule employs equal weighting of returns over n periods and therefore it is "tailored" to some specific duration of momentum. As a result, the MOM(n) rule works well only when the duration of momentum is comparable with the size of the averaging window. In contrast, the P-SMA(n) rule overweights the most recent returns and underweight the most distant returns; this weighting scheme allows one to almost fully exploit the short-term momentum and account for long-term momentum. As a result, the P-SMA(n) rule is able to exploit the momentum effect without knowing its precise duration (a similar discussion can be found in Hong and Satchell 2015).

Strictly speaking, our results say that the moving average trading strategy had some advantages (reasonable downside protection without a significant reduction in returns) in the past, even after accounting for such market frictions as transaction costs. A natural question that can be raised now is: Will the moving average strategy show the same advantages in the future as in the past? Whereas the past performance is not a guarantee of future performance, there are several reasons that advocate that the advantages are likely to persist in the future:

- The performance of the P-SMA(n) rule is robust to the choice of n, see the discussion above.
- Our results on the performance of the moving average trading strategy can be criticized, as a matter of fact, on the grounds that our out-of-sample tests

are not truly out-of-sample. Indeed, a truly out-of-sample test requires using a new set of rules or/and a new dataset. These requirements are not met in our tests: we used the existing set of rules and the dataset that overlaps to a large degree with the datasets used in many other studies. However, the superior performance of the 200-day (10-month) SMA rule was documented already by Gartley (1935). Afterwards, the superior performance of this rule after the period of the Great Depression was documented by Brock et al. (1992). After that, this rule delivered superior performance during the Dot-Com bubble crash of 2001–2002, see, for example, Faber (2007). Last but not least, this rule again outperformed the buy-and-hold strategy during the Global Financial Crisis of 2007–2008. The fact, that this rule keeps outperforming its passive counterpart each time after the superiority of this rule was documented in finance literature, is equivalent to a truly out-of-sample test of this rule. It is worth emphasizing, however, that this rule works mainly during a rather long bear market when prices decline steadily but not sharply; this rule does not work when prices suddenly drop as in October 1987.

- The momentum effect in stock prices is considered an "anomaly" in academic finance literature. The common criticism that can be raised in this regard is that "once an apparent anomaly is publicized, only too often it disappears or goes into reverse" (see, for example, Dimson and Marsh 1999). Indeed, when many traders try to profit from an existing anomaly, it can disappear or go into reverse. For example, when some type of stocks become popular because they perform better than the rest of the market, and when traders rash to buy these stocks, the return on these stocks, and hence the performance, deteriorates. However, there are some anomalies that can be strengthened when many traders want to profit from them. The momentum anomaly is an example of such anomaly. This is because when many traders sell (buy) the stocks when the prices go below (above) a 10-month SMA, this massive sale (purchase) only reinforces the downtrend (uptrend) in the stock prices.

- The momentum in financial asset prices is pervasive across a wide variety of investment universes, geographies, and even asset classes, see Moskowitz et al. (2012).

- Researchers advocate that the momentum effect has intuitive explanations grounded in strong behavioral arguments: initial under-reaction and delayed over-reaction. Under-reaction usually results from the slow diffusion of news, conservativeness, and the fact that price adjustment to new information takes some time. Over-reaction can be caused by positive feedback trading and over-confidence. Additional our own explanation for the momentum effect lies in the forecasting methodology and a self-fulfilling prophecy of some forecasts.[3] Specifically, traders usually forecast the future price direction by extrapolating the price trend in the recent past. When many traders strongly believe in their forecasts and start acting on them, their trades (selling or buying pressure) ultimately fulfill the "prophecy". Our explanation for the momentum effect can account for the fact that the duration of momentum varies over time. For example, in a calm market when prices trend steadily one can identify the direction of the price trend using just a few past prices. In contrast, in a turbulent market when prices fluctuate wildly it is difficult to identify the direction of a price trend using a few past prices and therefore one needs to use a longer past period. Consequently, our explanation predicts that the duration of momentum depends on the market volatility: the duration of momentum should be shorter (longer) when volatility is low (high). This property of momentum was observed by Kaufman (1995) who suggested using an Adaptive Moving Average where the size of the averaging window is directly proportional to the market volatility.

References

Balvers, R., Wu, Y., & Gilliland, E. (2000). Mean reversion across national stock markets and parametric contrarian investment strategies. *Journal of Finance, 55*(2), 745–772.

Box, G. E. P., & Jenkins, J. M. (1976). *Time series analysis: Forecasting and control.* San Francisco, CA: Holden-Day.

Brock, W., Lakonishok, J., & LeBaron, B. (1992). Simple technical trading rules and the stochastic properties of stock returns. *Journal of Finance, 47*(5), 1731–1764.

Dimson, E., & Marsh, P. (1999). Murphy's law and market anomalies. *Journal of Portfolio Management, 25*(2), 53–69.

Faber, M. T. (2007). A quantitative approach to tactical asset allocation. *Journal of Wealth Management, 9*(4), 69–79.

Fama, E., & French, K. (1988). Permanent and temporary components of stock prices. *Journal of Political Economy, 96*(2), 246–273.

Gartley, H. M. (1935). *Profits in the stock market.* Lambert Gann.

Hong, K. J., & Satchell, S. (2015). Time series momentum trading strategy and autocorrelation amplification. *Quantitative Finance, 15*(9), 1471–1487.

Jegadeesh, N. (1991). Seasonality in stock price mean reversion: Evidence from the U.S. and the U.K. *Journal of Finance, 46*(4), 1427–1444.

---

[3]A self-fulfilling prophecy is a prediction that directly or indirectly causes itself to become true, by the very terms of the prophecy itself, due to positive feedback between belief and behavior.

Jegadeesh, N., & Titman, S. (1993). Returns to buying winners and selling losers: Implications for stock market efficiency. *Journal of Finance, 48*(1), 65–91.

Kaufman, P. J. (1995). *Smarter trading: Improving performance in changing markets.* McGraw-Hill.

Kleinman, G. (2005). *Trading commodities and financial futures: A step-by-step guide to mastering the markets.* Pearson Education, Inc.

Moskowitz, T. J., Ooi, Y. H., & Pedersen, L. H. (2012). Time series momentum. *Journal of Financial Economics, 104*(2), 228–250.

Index

A

Asset allocation puzzles, 200
Asset classes, 203
Auto-regressive model, 268
Average lag time
    double exponential smoothing, 38
    exponential moving average, 31
    general weighted moving average,
        13
    linear moving average, 26
    simple moving average, 24
    triangular moving average, 36
    triple exponential smoothing, 38

B

Back test, 130
    bond markets, 235
    commodity markets, 251
    currency markets, 241
    rolling, 169
    S&P Composite index, daily data,
        189
    S&P Composite index, monthly
        data, 165
    stock markets, 226
Bootstrap
    block bootstrap, 124
    standard bootstrap, 123
    stationary block bootstrap, 125
Bretton Woods Agreements, 239

Bull and bear markets, 150
    dating, 150
    definition, 150
    grains, 256
    long-term bonds, 233
    precious metals, 249
    S&P Composite index, 151
    US dollar, 241

C

Capital allocation problem, 113, 198,
    206
Commodity, 248
Commodity Trading Advisor (CTA),
    250
Cost of carry, 250

D

Data-mining, 130
Data-mining bias, 130
Dendrogram, 170
Double exponential moving average
    (DEMA), 43
Double exponential smoothing, 37
Dow Jones Industrial Average (DJIA)
    index, 224
Downside standard deviation, 118
Dow Theory, 258
Drawdown, 173

© The Author(s) 2017
V. Zakamulin, *Market Timing with Moving Averages*, New Developments
in Quantitative Trading and Investment, DOI 10.1007/978-3-319-60970-6

E

Equivalence of two technical indicators, 72
Equivalent technical indicator
    momentum rule, 73
    moving average change of direction rule, 80
    moving average convergence/divergence rule, 89
    moving average crossover rule, 83
    price minus moving average rule, 75
Exchange rate, 238
Exchange-rate regime, 238
    fixed rates, 238
    floating rates, 239
Expected loss if loss occurs, 208, 209, 215, 262
Exponential moving average (EMA), 30
    average lag time, 31, 51
    Herfindahl index, 33, 51
    smoothness, 34

F

Forward test, 133
    bond markets, 236
    commodity markets, 252
    currency markets, 243
    S&P Composite index, daily data, 193
    S&P Composite index, monthly data, 172, 179
    stock markets, 229
Fundamental analysis, 3

G

General weighted moving average, 11
Gold exchange standard, 239
Gold standard, 238
Growth stocks, 225

H

Herfindahl index, 19
Hull moving average (HMA), 45
Hypothesis testing, 120

I

In-sample test, 130
Intermediate-term bond index, 233
Investor type
    aggressive, 201
    conservative, 201, 271
    moderate, 201, 271

K

Kurtosis, 173

L

Large cap stocks, 225
Linearity property of moving averages, 12
Linear moving average (LMA), 25
    average lag time, 26, 50
    Herfindahl index, 28, 50
    smoothness, 28
Long-term bond index, 233

M

Margin of safety investment principle, 202
Mean excess return, 113
Mean-variance utility, 114, 197
Momentum rule, 56
Moving average, 4
    alternative representation, 14
    centered, 4
    double exponential (DEMA), 43
    double exponential smoothing, 37
    exponential (EMA), 30
    general weighted, 11
    Hull (HMA), 45
    as a linear operator, 12
    linear (LMA), 25

price-change weighting function, 15
price weighting function, 12, 15
right-aligned, 6
simple (SMA), 23
triangular (TMA), 35
triple exponential (TEMA), 43
triple exponential smoothing, 37
zero lag exponential (ZLEMA), 39
Moving average change of direction rule, 57
Moving average convergence/divergence rule, 63
Moving average crossover rule, 60
Moving average envelope, 66
Moving average ribbon, 62

O
Out-of-sample test, 133, 135
Outperformance
  definition, 120
  null hypothesis, 120

P
Parametric test
  mean excess return, 122
  Sharpe ratio, 122
Portfolio insurance strategy, 210
Portfolio performance measure
  estimation, 119
  mean excess return, 113
  Sharpe ratio, 115
  Sortino ratio, 118
60/40 portfolio of stocks and bonds, 208, 210, 215
Price minus moving average rule, 58
Price-change weighting function, 15
  momentum rule, 74
  moving average change of direction rule, 81

moving average convergence/divergence rule, 89
moving average crossover rule, 85
price minus moving average rule, 76
Price weighting function, 15
Probability of loss, 203, 204, 207–209, 215, 262

R
Random Walk, 16, 19, 75, 258, 269
Returns to a trading strategy
  with short sales, 109
  without short sales, 108
Risk-free (safe) asset, 111, 201
Risk-free rate of return, 108, 144, 233, 240, 250
Risk measure
  downside standard deviation, 118
  drawdown, 173
  expected loss, 208
  probability of loss, 208
  standard deviation, 114

S
Secular bull and bear markets, 233
Self-fulfilling prophecy, 273
Sharpe ratio, 115, 156–158, 160, 163, 165–167, 173, 180, 184, 185, 189, 196, 224, 226, 230
Short sale strategy
  commodity markets, 255
  currency markets, 246
  stock markets, 160
Signal-to-noise ratio, 194, 242
Simple matching coefficient, 186
Simple moving average (SMA), 23
  average lag time, 24, 49
  Herfindahl index, 24
  smoothness, 24
Skewness, 173
Small cap stocks, 226

Smoothness
   exponential moving average, 33
   general weighted moving average,
     19
   linear moving average, 28
   simple moving average, 24
Sortino ratio, 118, 156, 157
Standard and Poor's Composite index,
   143, 144
Standard and Poor's 500 index, 143
Structural break analysis, 146, 216

T
Technical analysis, 3
Technical trading indicator, 55
   alternative representation, 91, 266
   momentum rule, 56
   moving average change of direction
     rule, 57
   moving average conver-
     gence/divergence rule,
     63
   moving average crossover rule, 60
   as a predictive linear model, 268
   price minus moving average rule,
     59
Testing trading rules
   back (in-sample) test, 130
   forward (out-of-sample) test, 133
   walk-forward test, 135

Trading signal generation, 56
   moving average envelope, 67
Transaction Costs, 105
   bid-ask spread, 105
   bond markets, 107
   brokerage fees (commissions), 105
   for large investors, 105
   market impact, 106
   for small investors, 106
   stock markets, 107
Treasury Bill rate, 144
Trend following, 4
Triangular moving average (TMA),
   35
   average lag time, 36
Triple exponential moving average
   (TEMA), 43
Triple exponential smoothing, 37

V
Value stocks, 225

W
Walk-forward test, 135
Whipsaw trades, 65

Z
Zero lag exponential moving average
   (ZLEMA), 39

Printed by Printforce, the Netherlands